"十四五"高等职业教育人工智能技术应用系列教材

RENGONG ZHINENG
TUXIANG SHIBIE XIANGMU SHIJIAN

人工智能
图像识别项目实践

唐 林 吴丰盛 ◎ 主 编
曹宏钦 刘 成 张 霖 ◎ 副主编

中国铁道出版社有限公司
CHINA RAILWAY PUBLISHING HOUSE CO., LTD.

内 容 简 介

本书按项目任务式编写，以解决复杂的项目问题为核心，围绕着如何使用人工智能图像识别技术展开，由浅入深、层层递进，提升读者的综合思维能力和应用能力。

本书共6个项目，主要内容包括农业病虫害图像数据增强、动漫自动设计图像标注、宠物管理猫狗检测、自动驾驶行人检测、智慧社区交通工具检测、节能洗车房车牌识别。

本书配套的源代码以及相关数字化资源可在中国铁道出版社有限公司网站及中育数据官网资源中心栏目下载。

本书适合作为高等职业院校人工智能课程的教材，也可作为普通读者学习图像识别技术应用的参考用书。

图书在版编目(CIP)数据

人工智能图像识别项目实践/唐林，吴丰盛主编．—北京：中国铁道出版社有限公司，2022.1（2023.12重印）
"十四五"高等职业教育人工智能技术应用系列教材
ISBN 978-7-113-28709-2

Ⅰ.①人… Ⅱ.①唐… ②吴… Ⅲ.①人工智能-算法-应用-图像识别-高等职业教育-教材 Ⅳ.①TP391.413

中国版本图书馆CIP数据核字(2021)第261971号

书　　名：人工智能图像识别项目实践
作　　者：唐　林　吴丰盛

策划编辑：祁　云　　　　　　　编辑部电话：(010)63549458
责任编辑：祁　云　包　宁
封面设计：尚明龙
责任校对：焦桂荣
责任印制：樊启鹏

出版发行：中国铁道出版社有限公司(100054，北京市西城区右安门西街8号)
网　　址：http://www.tdpress.com/51eds/
印　　刷：三河市国英印务有限公司
版　　次：2022年1月第1版　2023年12月第2次印刷
开　　本：787 mm×1 092 mm　1/16　印张：10　字数：256千
书　　号：ISBN 978-7-113-28709-2
定　　价：35.00元

版权所有　侵权必究

凡购买铁道版图书，如有印制质量问题，请与本社教材图书营销部联系调换。电话：(010)63550836
打击盗版举报电话：(010)63549461

前　言

"学习的内容是工作,通过工作实现学习",这是教育界一位资深专家、博士生导师对"职业教育工学结合一体化课程"提出的观点,也是我们创作本书的初衷。希望读者通过编者的文字感受到真实场景,通过完成项目任务提升综合能力。

人工智能的迅速发展将深刻改变人类社会生活、改变世界。过去几年,人工智能已经让一部分行业产生了变革,从国家发展层面来看,必将带来更大的机遇。中国正处于全面建成小康社会的决胜阶段,人口老龄化、资源环境约束等挑战依然严峻,人工智能在教育、医疗、养老、环境保护、城市运行、司法服务等领域广泛应用,将极大地提高公共服务精准化水平,全面提升人民生活品质。

每个人都能够在这场人工智能的变革中找到自己的角色,不论学历、不论行业。随着人工智能算法的工具化,使用人工智能技术的人也必将从科学家扩展到普通人,让我们举一个目标检测的例子来解释说明。传统目标检测算法需要使用者具有非常高的数学功底和编程能力,要找到既懂业务又懂人工智能应用开发的人非常难;进入算法工具化时代之后,基于深度学习的目标检测不再需要定制算法,典型的应用开发流程变成了收集图片、标注图片、训练模型和部署模型,人们可以花更多的时间在分析业务需求上。只要你是行业的专家,就可以成为人工智能技术的真正驾驭者。

让读者在更短时间内成为人工智能图像识别技术的驾驭者。本书编者长期工作在教学科研、企业项目一线,以解决复杂的项目问题为核心,围绕着如何使用人工智能图像识别技术,由浅入深、层层递进,提升读者的综合思维能力和应用能力。第一个项目案例来自农业科学研究院,解决病虫害的问题,学习图像增强的技能;第二个项目案例来自动漫设计公司,解决人物自动设计的问题,学习图像标注的技能;第三、第五个项目案例来自转型中的系统集成公司,解决宠物店管理和社区停车难的问题,学习目标检测的技能;第四、第六个项目案例来自传统的制造型企业,它们都通过人工智能技术帮助企业抓住了新的机遇,分别解决了传感器感知行人的问题和洗车房识别车牌的问题,同时增加了对目标检测技术应用的熟练程度。

本书适合以掌握人工智能技术应用技能为学习目的的读者，特别是高职院校的学生使用。帮助读者树立业务思维模式，坚定地站在业务的角度，把人工智能技术当成工具，综合地加以运用。书中涉及的相关工具软件，编者团队通过互联网搭建了演示系统，提供免费共享账号，有需要的读者可以扫码添加微信索取。

本书由唐林、吴丰盛任主编，由曹宏钦、刘成、张霖任副主编，和中育数据研发团队共同编写完成。由于编者水平有限，加之时间仓促，书中难免存在疏漏和不足之处，恳请读者批评指正。

编 者

2021 年 10 月

目 录

项目1　农业病虫害图像数据增强 …… 1

项目背景 …… 1
任务1　工程环境准备 …… 1
任务2　图像水平翻转 …… 3
任务3　图像旋转 …… 4
任务4　图像缩放 …… 6
任务5　图像高斯噪声 …… 7
任务6　图像高斯模糊 …… 8
任务7　图像转灰度 …… 10
任务8　图像增亮度 …… 11
任务9　图像增色调 …… 12
任务10　图像增饱和度 …… 14
项目小结 …… 15

项目2　动漫自动设计图像标注 …… 16

项目背景 …… 16
任务1　工程环境准备 …… 16
任务2　创建动漫人脸标签 …… 17
任务3　创建标注任务 …… 18
任务4　标注动漫人脸图片 …… 19
任务5　完成标注任务 …… 22
任务6　导出数据集 …… 23
项目小结 …… 25

项目3　宠物管理猫狗检测 …… 26

项目背景 …… 26
任务1　数据准备 …… 26
任务2　工程环境准备 …… 27
任务3　猫狗图片数据标注 …… 30
任务4　猫狗检测模型训练 …… 34
任务5　猫狗检测模型评估 …… 44
任务6　猫狗检测模型测试 …… 47

任务7　猫狗检测模型部署 ………………………………………………… 52
　　项目小结 ……………………………………………………………………… 58

项目4　自动驾驶行人检测 ……………………………………………… 59
　　项目背景 ……………………………………………………………………… 59
　　任务1　数据准备 …………………………………………………………… 59
　　任务2　工程环境准备 ……………………………………………………… 60
　　任务3　行人图片数据标注 ………………………………………………… 63
　　任务4　行人检测模型训练 ………………………………………………… 67
　　任务5　行人检测模型评估 ………………………………………………… 77
　　任务6　行人检测模型测试 ………………………………………………… 80
　　任务7　行人检测模型部署 ………………………………………………… 85
　　项目小结 ……………………………………………………………………… 90

项目5　智慧社区交通工具检测 ………………………………………… 91
　　项目背景 ……………………………………………………………………… 91
　　任务1　数据准备 …………………………………………………………… 91
　　任务2　工程环境准备 ……………………………………………………… 92
　　任务3　交通工具图片数据标注 …………………………………………… 95
　　任务4　交通工具检测模型训练 …………………………………………… 99
　　任务5　交通工具检测模型评估 …………………………………………… 108
　　任务6　交通工具检测模型测试 …………………………………………… 112
　　任务7　交通工具检测模型部署 …………………………………………… 116
　　项目小结 ……………………………………………………………………… 122

项目6　节能洗车房车牌识别 …………………………………………… 123
　　项目背景 ……………………………………………………………………… 123
　　任务1　数据准备 …………………………………………………………… 123
　　任务2　工程环境准备 ……………………………………………………… 124
　　任务3　车牌图片数据标注 ………………………………………………… 127
　　任务4　车牌识别模型训练 ………………………………………………… 131
　　任务5　车牌识别模型评估 ………………………………………………… 140
　　任务6　车牌识别模型测试 ………………………………………………… 144
　　任务7　车牌识别模型部署 ………………………………………………… 148
　　项目小结 ……………………………………………………………………… 154

项目 1

农业病虫害图像数据增强

项目背景

热爱家乡的你从学校毕业后,回到了所在市的一家农业科学研究所工作。市所属的一个县为响应国家精准扶贫号召,引入了木薯种植产业,质优物美的木薯淀粉成为乡亲们致富的出路。你所在的团队接到了一个关于木薯叶病虫害检测的项目,需要通过图像识别检测出病虫害的种类,以此来帮助种植户快速找到问题原因,对症下药,及时采取应对措施。项目启动会上,领导告诉大家:如果项目实施效果满意,会推广到全省甚至其他省,帮助更多的种植户解决类似的问题,所以请大家务必重视这个项目的实用性和通用性。工作开始后,团队从现场采集到了一部分照片,种植户也提供了一部分病虫害的照片,大约有100张。项目经理要求提供1 000张图片用于数据标注和模型训练,这项工作需要你完成,请尽快完成数据扩充工作。

提示:深层神经网络一般都需要一定数量级的训练数据才能获得理想的结果,但是很多实际项目,都难以有充足的数据,要保证完美地完成任务,有两件事情需要做好:①寻找更多数据;②充分利用已有数据进行数据增强。在计算机视觉中,典型的数据增强方法有翻转、旋转、缩放、调整灰度、调整亮度、调整色调和饱和度等。数据增强的好处是提高模型鲁棒性(又称健壮性),避免过拟合。本项目提供了九种常见的图像数据增强方法,下面一起完成项目任务。

任务 1　工程环境准备

任务描述

准备数据增强对应的工程环境。本任务将创建项目工程环境,并把相关图像数据导入工程环境。

步骤1 创建工程目录

在开发环境中为本项目创建工程目录,在终端命令行窗口中执行以下操作:

* 注意需要把命令中的地址换成对应的资源平台地址。

```
$ mkdir ~/projects/unit1
$ mkdir ~/projects/unit1/img
$ cd ~/projects/unit1/img
$ wget http://172.16.33.72/dataset/leaf.tar.gz
$ tar zxvf leaf.tar.gz
$ rm leaf.tar.gz
```

刷新目录后打开一张图片查看,结果如图1.1所示。

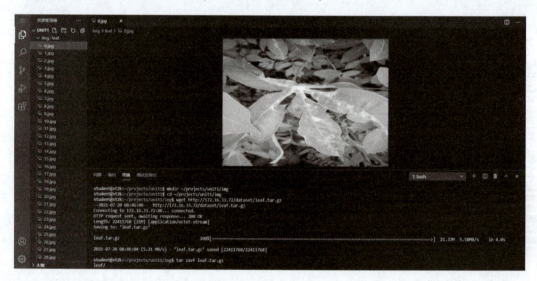

图1.1 待处理的木薯叶样例图

步骤2 创建开发环境

创建名为unit1的虚拟环境,在Python 3.6版本中执行如下操作:

```
$ conda create -n unit1 python=3.6
```

输入y继续完成,然后执行如下操作激活开发环境:

```
$ conda activate unit1
```

在开发环境中执行如下操作,安装opencv库和imgaug库并查看结果,如图1.2所示。

```
$ pip install opencv-python imgaug
```

项目 1　农业病虫害图像数据增强

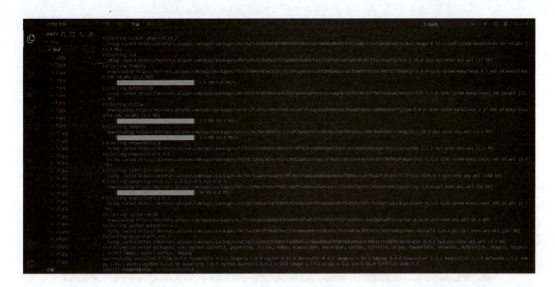

图 1.2　创建开发环境

任务 2　图像水平翻转

任务描述

执行水平翻转处理操作。本任务中将对原始图片进行水平翻转处理,输出后另存为新的图片文件。

任务操作

步骤 1　创建处理文件

在开发环境中打开 /home/student/projects/unit1/ 目录,创建 u1_2.py 图片处理文件。

将以下内容写到 u1_2.py 文件中。主要内容包括导入 opencv、imgaug 库,利用 opencv 库读取输入图片文件,利用 imgaug 库创建一个水平翻转序列增强器,利用 imgaug 库对输入图片进行操作,利用 opencv 库输出图片文件。

```
import cv2
import imgaug.augmenters as iaa

input_img = cv2.imread("./img/leaf/0.jpg")
seq = iaa.Sequential([iaa.Fliplr(1)])
output_img = seq.augment_image(input_img)
cv2.imwrite("./img/0_2.jpg",output_img)
print("已完成输入图片的水平翻转处理!")
```

步骤 2　处理图片

运行程序 u1_2.py,执行以下操作,运行结果如图 1.3 所示,查看输出图片,如图 1.4 所示。

```
$ conda activate unit1
$ python u1_2.py
```

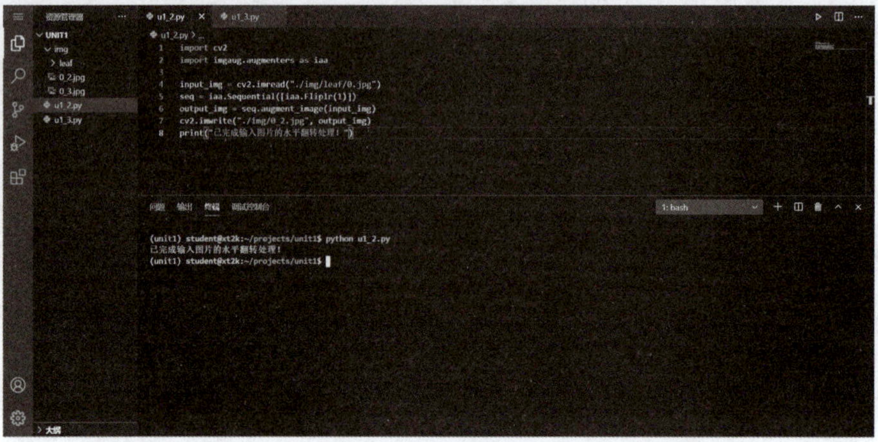

图1.3 图像水平翻转任务

图1.4 水平翻转图片

任务3　图像旋转

任务描述

执行旋转处理操作。本任务中将对原始图片进行旋转处理,输出后另存为新的图片文件。

任务操作

步骤1　创建处理文件

在开发环境中打开/home/student/projects/unit1/目录,创建 u1_3.py 图片处理文件。

项目 1　农业病虫害图像数据增强

将以下内容写到 u1_3.py 文件中。主要操作包括导入 opencv、imgaug 库,利用 opencv 库读取输入图片文件,利用 imgaug 库创建一个旋转序列增强器,利用 imgaug 库对输入图片进行操作,利用 opencv 库输出图片文件。

```
import cv2
import imgaug.augmenters as iaa

input_img = cv2.imread("./img/leaf/0.jpg")
seq = iaa.Sequential([iaa.Affine(rotate=(-30,30))])
output_img = seq.augment_image(input_img)
cv2.imwrite("./img/0_3.jpg",output_img)
print("已完成输入图片的旋转处理!")
```

步骤 2　处理图片

运行程序 u1_3.py,执行以下操作,运行结果如图 1.5 所示,查看输出图片,如图 1.6 所示。

```
$ python u1_3.py
```

图 1.5　图像旋转任务

图 1.6　图片旋转 30°

任务4　图像缩放

任务描述

执行缩放处理操作。本任务中将对原始图片进行缩放处理,输出后另存为新的图片文件。

任务操作

步骤1　创建处理文件

在开发环境中打开/home/student/projects/unit1/目录,创建 u1_4.py 图片处理文件。

将以下内容写到 u1_4.py 文件中。主要操作包括导入 opencv、imgaug 库,利用 opencv 库读取输入图片文件,利用 imgaug 库创建一个缩放(350×350像素正方形)序列增强器,利用 imgaug 库对输入图片进行操作,利用 opencv 库输出图片文件。

```
import cv2
import imgaug.augmenters as iaa

input_img = cv2.imread("./img/leaf/0.jpg")
seq = iaa.Sequential([iaa.Resize(size=[350,350],interpolation='nearest')])
output_img = seq.augment_image(input_img)
cv2.imwrite("./img/0_4.jpg",output_img)
print("已完成输入图片的缩放处理!")
```

步骤2　处理图片

运行程序 u1_4.py,执行以下操作,运行结果如图1.7所示,查看输出图片,如图1.8所示。

图1.7　图像缩放任务

```
$ python u1_4.py
```

图 1.8　缩放为 350×350 像素尺寸的图片

任务 5　图像高斯噪声

任务描述

执行高斯噪声处理操作。本任务中将对原始图片进行高斯噪声处理，输出后另存为新的图片文件。

任务操作

步骤 1　创建处理文件

在开发环境中打开 /home/student/projects/unit1/ 目录，创建 u1_5.py 图片处理文件。

将以下内容写到 u1_5.py 文件中。主要操作包括导入 opencv、imgaug 库，利用 opencv 库读取输入图片文件，利用 imgaug 库创建一个高斯噪声序列增强器，利用 imgaug 库对输入图片进行操作，利用 opencv 库输出图片文件。

```python
import cv2
import imgaug.augmenters as iaa

input_img = cv2.imread("./img/leaf/0.jpg")
seq = iaa.Sequential([iaa.AdditiveGaussianNoise(scale=(0,0.05 * 255))])
output_img = seq.augment_image(input_img)
cv2.imwrite("./img/0_5.jpg",output_img)
print("已完成输入图片的高斯噪声处理!")
```

步骤 2　处理图片

运行程序 u1_5.py，执行以下操作，运行结果如图 1.9 所示，查看输出图片，如图 1.10 所示。

```
$ python u1_5.py
```

图 1.9　图像高斯噪声任务

图 1.10　加入高斯噪声的图片

任务 6　图像高斯模糊

任务描述

执行高斯模糊处理操作。本任务中将对原始图片进行高斯模糊处理,输出后另存为新的图片文件。

任务操作

步骤 1　创建处理文件

在开发环境中打开 /home/student/projects/unit1/ 目录,创建 u1_6.py 图片处理文件。

将以下内容写到 u1_6.py 文件中。主要操作包括导入 opencv、imgaug 库,利用 opencv 库读取输入图片文件,利用 imgaug 库创建一个高斯模糊序列增强器,利用 imgaug 库对输入图片进行操作,利用 opencv 库输出图片文件。

```
import cv2
import imgaug.augmenters as iaa

input_img = cv2.imread("./img/leaf/0.jpg")
seq = iaa.Sequential([iaa.GaussianBlur(sigma = (0.0,3.0))])
output_img = seq.augment_image(input_img)
cv2.imwrite("./img/0_6.jpg",output_img)
print("已完成输入图片的高斯模糊处理!")
```

步骤 2　处理图片

运行程序 u1_6.py,执行以下操作,运行结果如图 1.11 所示,查看输出图片,如图 1.12 所示。

```
$ python u1_6.py
```

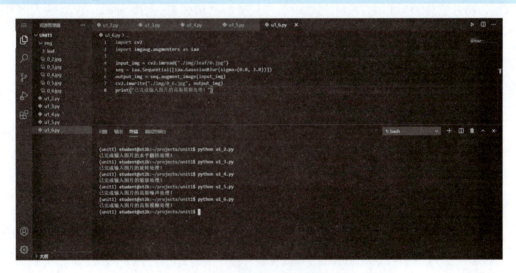

图 1.11　图像高斯模糊任务

图 1.12　加入高斯模糊的图片

任务 7　图像转灰度

任务描述

执行转灰度处理操作。本任务中将对原始图片进行转灰度处理,输出后另存为新的图片文件。

任务操作

步骤 1　创建处理文件

在开发环境中打开/home/student/projects/unit1/目录,创建 u1_7.py 图片处理文件。

将以下内容写到 u1_7.py 文件中。主要操作包括导入 opencv、imgaug 库,利用 opencv 库读取输入图片文件,利用 imgaug 库创建一个转灰度序列增强器,利用 imgaug 库对输入图片进行操作,利用 opencv 库输出图片文件。

```python
import cv2
import imgaug.augmenters as iaa

input_img = cv2.imread("./img/leaf/0.jpg")
seq = iaa.Sequential([iaa.Grayscale(alpha=(0.0,1.0))])
output_img = seq.augment_image(input_img)
cv2.imwrite("./img/0_7.jpg",output_img)
print("已完成输入图片的转灰度处理!")
```

步骤 2　处理图片

运行程序 u1_7.py,执行以下操作,运行结果如图 1.13 所示,查看输出图片,如图 1.14 所示。

```
$ python u1_7.py
```

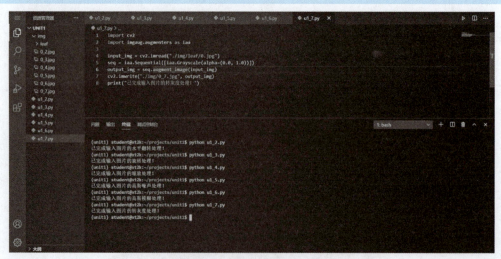

图 1.13　图像转灰度任务

项目 1　农业病虫害图像数据增强

图 1.14　转灰度的图片

任务 8　图像增亮度

任务描述

执行增亮度处理操作。本任务中将对原始图片进行增亮度处理,输出后另存为新的图片文件。

任务操作

步骤 1　创建处理文件

在开发环境中打开/home/student/projects/unit1/目录,创建 u1_8.py 图片处理文件。

将以下内容写到 u1_8.py 文件中。主要操作包括导入 opencv、imgaug 库,利用 opencv 库读取输入图片文件,利用 imgaug 库创建一个增亮度序列增强器,利用 imgaug 库对输入图片进行操作,利用 opencv 库输出图片文件。

```
import cv2
import imgaug.augmenters as iaa

input_img = cv2.imread("./img/leaf/0.jpg")
seq = iaa.Sequential([iaa.AddToBrightness((-50,50))])
output_img = seq.augment_image(input_img)
cv2.imwrite("./img/0_8.jpg",output_img)
print("已完成输入图片的增亮度处理!")
```

步骤 2　处理图片

运行程序 u1_8.py,执行以下操作,运行结果如图 1.15 所示,查看输出图片,如图 1.16 所示。

```
$ python u1_8.py
```

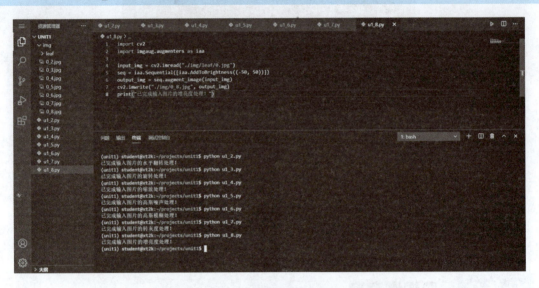

图 1.15 图像增亮度任务

图 1.16 增亮度的图片

任务9　图像增色调

任务描述

执行增色调处理操作。本任务中将对原始图片进行增色调处理,输出后另存为新的图片文件。

项目 1　农业病虫害图像数据增强

 任务操作

步骤 1　创建处理文件

在开发环境中打开 /home/student/projects/unit1/ 目录,创建 u1_9.py 图片处理文件。

将以下内容写到 u1_9.py 文件中。主要操作包括导入 opencv、imgaug 库,利用 opencv 库读取输入图片文件,利用 imgaug 库创建一个增色调序列增强器,利用 imgaug 库对输入图片进行操作,利用 opencv 库输出图片文件。

```
import cv2
import imgaug.augmenters as iaa

input_img = cv2.imread("./img/leaf/0.jpg")
seq = iaa.Sequential([iaa.AddToHue((-150,150))])
output_img = seq.augment_image(input_img)
cv2.imwrite("./img/0_9.jpg",output_img)
print("已完成输入图片的增色调处理!")
```

步骤 2　处理图片

运行程序 u1_9.py,执行以下操作,运行结果如图 1.17 所示,查看输出图片,如图 1.18 所示。

```
$ python u1_9.py
```

图 1.17　图像增色调任务

13

图 1.18 增色调的图片

任务 10　图像增饱和度

任务描述

执行增饱和度处理操作。本任务中将对原始图片进行增饱和度处理,输出后另存为新的图片文件。

任务操作

步骤 1　创建处理文件

在开发环境中打开 /home/student/projects/unit1/ 目录,创建 u1_10.py 图片处理文件。

将以下内容写到 u1_10.py 文件中。主要操作包括导入 opencv、imgaug 库,利用 opencv 库读取输入图片文件,利用 imgaug 库创建一个增饱和度序列增强器,利用 imgaug 库对输入图片进行操作,利用 opencv 库输出图片文件。

```
import cv2
import imgaug.augmenters as iaa

input_img = cv2.imread("./img/leaf/0.jpg")
seq = iaa.Sequential([iaa.AddToSaturation((-150,150))])
output_img = seq.augment_image(input_img)
cv2.imwrite("./img/0_10.jpg",output_img)
print("已完成输入图片的增饱和度处理!")
```

步骤 2　处理图片

运行程序 u1_10.py,执行以下操作,运行结果如图 1.19 所示,查看输出图片,如图 1.20 所示。

```
$ python u1_10.py
```

图 1.19 图像增饱和度任务

图 1.20 增饱和度的图片

项目小结

祝贺你和你的团队,木薯叶病虫害检测项目的图片顺利从 100 张扩充到了 1 000 张,为项目的后续工作打下了良好的基础。在本项目的任务中用到的 imgaug 库是一个已经封装好的用来进行图像数据增强的 Python 库,可以将输入图片转换成多种输出图片。

多学一点:为了防止过拟合,数据增强应运而生。随着神经网络层数的增加,模型需要学习的参数也会随之增加,这样就更容易产生过拟合,也就是模型对训练数据识别度很高但是对于测试数据的准确率却很低。除了数据增强,还有正则项/dropout 等方式可以防止过拟合。数据增强可以分为有监督的数据增强和无监督的数据增强方法,其中有监督的数据增强又可以分为单样本数据增强和多样本数据增强方法,无监督的数据增强分为生成新的数据和学习增强策略两个方向。祝愿你在未来的学习中掌握更多的技能,在实际工作中选择最有效的防止过拟合的方法,成为一名优秀的工程师。

项目 2

动漫自动设计图像标注

项目背景

热爱生活的你毕业后进入了一家动漫设计公司。虽然这是一家创业公司，但是却有着一个数百万卡通形象的数据库，设计师只需要输入某个经典卡通人物的名字，马上就能得到关于这个卡通的图片和视频，以此帮助设计师寻找灵感。你所在的团队接到了一个新任务，目标是帮助设计师"快速"锁定原型。主要方式是：设计师提供卡通人物的图片，系统自动创作类似的卡通形象，设计师只需根据新人物的特点进行针对性的调整设计即可。项目启动会上，领导告诉大家：这是一个全新的尝试，不管结果怎么样，希望大家齐心协力、努力攻关。工作开始后，项目团队被分成两个小组，你所在的小组负责实现卡通形象的识别，找到相关特征值；另一个小组负责把特征值进行调整，生成新的卡通形象。

提示：计算机视觉是人工智能的一个重要方向，是人机交互的基础之一，它负责解释接收到的图片和视频数据。图像标注在计算机视觉中起着至关重要的作用，通过标注可以让计算机具备以下能力：目标检测、目标分割、图像分类、姿态预测和关键点识别等。本项目介绍常用的图像标注方法，包括矩形框和关键点，其中矩形框主要用于训练计算机的目标检测能力，关键点主要用于关键特征识别，下面一起完成项目任务。

任务 1　工程环境准备

任务描述

准备数据标注对应的工程环境。本任务中将创建项目工程环境，并把需要标注的数据放到 DATA 子目录中。

项目 2　动漫自动设计图像标注

创建工程目录

在开发环境中为本项目创建工程目录,在终端命令行窗口中执行以下操作。

* 注意需要把命令中的地址修改成对应的资源平台地址。

```
$ cd ~/data
$ wget http://172.16.33.72/dataset/cartoon_face.tar.gz
$ tar zxvf cartoon_face.tar.gz
$ rm zxvf cartoon_face.tar.gz
```

刷新目录,打开 0.jpg 图片查看,结果如图 2.1 所示。

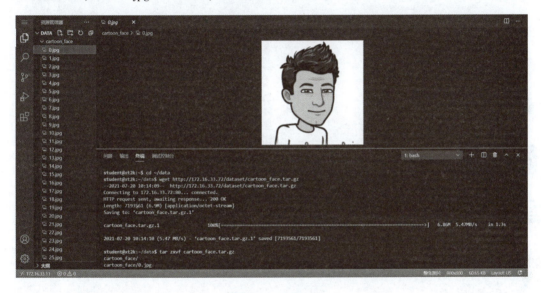

图 2.1　查看图片

任务 2　创建动漫人脸标签

创建标注标签。在本任务中将首先创建一个标注项目,然后创建 8 类标注标签。

步骤 1　创建标注项目

在图像标注工具中,创建名称为"动漫自动设计图像标注"的标注项目。

步骤 2　添加标注标签

(1) 添加 8 类标签,分别是脸框、脸颊、嘴唇、鼻子、左眼、右眼、左眉、右眉,如图 2.2 所示。
(2) 注意设置为不同的颜色标签以示区分。

17

(3)单击"提交"按钮完成标签的创建。

图 2.2 创建标签

任务 3　创建标注任务

任务描述

在本任务中将创建标注任务,注意一个项目可以有多个任务,比如训练集标注任务、验证集标注任务等,在这里仅创建一个"动漫人脸标注"任务。

任务操作

步骤 1　创建标注任务

(1)创建一个标注任务,名称为"动漫人脸标注"。
(2)项目选择上一节中完成的"动漫自动设计图像标注"。

步骤 2　选择数据文件

(1)选择"连接共享文件",如图 2.3 所示。
(2)选择本项目任务 1 中生成的数据子目录。
(3)单击"提交"按钮完成标签的创建。

项目 2　动漫自动设计图像标注

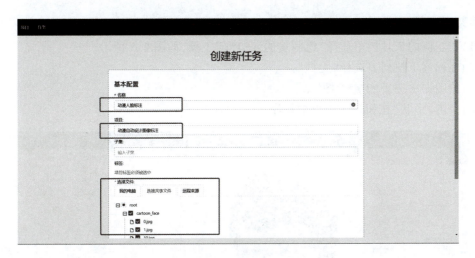

图 2.3　创建任务

任务 4　标注动漫人脸图片

对数据图片进行标注。在本任务中将使用矩形框标注和关键点标注方法，对一张数据图片进行标注，其中脸框标签用矩形框方法标注，其他 7 类标签用关键点方法标注。

任务操作

步骤 1　进入标注工作区

打开"动漫人脸标注"任务，单击左下方的"作业#"进入标注作业，如图 2.4 所示。

＊注意，"作业"旁边的 0～99 帧代表这个任务中一共有 100 张图片等待标注。

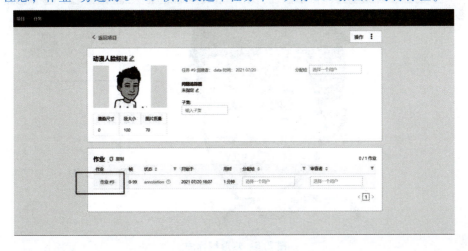

图 2.4　进入标注作业

步骤2　选择矩形框工具

（1）在左侧的工具栏中选择"绘制新的四边形"工具，如图2.5所示。
（2）在标签下拉列表选择"脸框"。
（3）绘图方法选择"2点绘图"。
（4）单击"形状"按钮进入矩形框标注。

图2.5　选择矩形框工具

步骤3　矩形框标注

（1）在卡通人像脸部单击并拉出一个矩形，然后再次单击或按【N】键完成标注，如图2.6所示。
（2）用鼠标放大图像，确保四个边缘与脸部图像贴合。
（3）"脸框"标注完成后，单击"锁"按钮，避免后面的误操作。

图2.6　矩形框标注

项目 2　动漫自动设计图像标注

步骤 4　选择关键点工具

（1）在左侧的工具栏中选择"绘制新的点"工具，如图 2.7 所示。
（2）在标签下拉列表中选择"脸颊"。
（3）点的数量输入"3"。
（4）单击"形状"按钮进入关键点标注。

图 2.7　选择关键点工具

步骤 5　关键点标注

（1）在卡通人像的脸部单击分别点出 3 个点，然后再次单击或按【N】键完成标注，如图 2.8 所示。
（2）"脸颊"标注完成后单击"锁"按钮，避免后面的误操作。

图 2.8　关键点标注

步骤6 完成标注

重复步骤4和5,完成以下标签的标注,如图2.9所示。

(1)关键点标注"嘴唇",点的数量为"3"。

(2)关键点标注"鼻子",点的数量为"3"。

(3)关键点标注"左眼",点的数量为"3"。

(4)关键点标注"右眼",点的数量为"3"。

(5)关键点标注"左眉",点的数量为"3"。

(6)关键点标注"右眉",点的数量为"3"。

图2.9 完成标注

任务5 完成标注任务

任务描述

完成标注任务。在本任务将通过下一帧、保存等操作,完成所有数据的标注工作。

任务操作

步骤1 标注下一张图

(1)在上方的工具栏中单击"下一帧"按钮,如图2.10所示。

(2)重复本项目任务4中的矩形框标注、关键点标注。

步骤2 保存标注

在上方的工具栏中单击"保存"按钮,避免工作丢失。

项目 2　动漫自动设计图像标注

图 2.10　下一帧和保存操作

任务 6　导出数据集

任务描述

将标注好的数据导出。在本任务将把标注好的数据导出，解压缩后给团队其他同事训练或获取中间结果使用。

任务操作

 导出标注文件

（1）选择"菜单"→"导出为数据集"→"CVAT for images 1.1"命令，如图 2.11 所示。

图 2.11　导出标注文件

(2)导出的压缩包文件保存在浏览器默认下载目录中;解压缩后打开 xml 文件,如图 2.12 所示,可以用来查看标注数据或做中间转换处理。

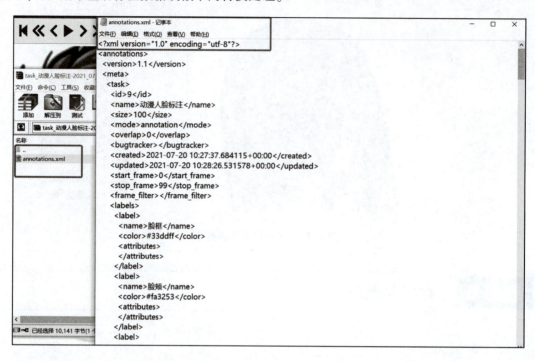

图 2.12　查看生成的 xml 文件

步骤 2　导出为数据集文件

(1)选择"菜单"→"导出为数据集"→"TFRecord 1.0"命令,如图 2.13 所示。

图 2.13　导出数据集

（2）导出的压缩包文件保存在浏览器默认下载目录中；解压缩后打开标签映射文件，如图 2.14 所示，此文件将用来训练模型。

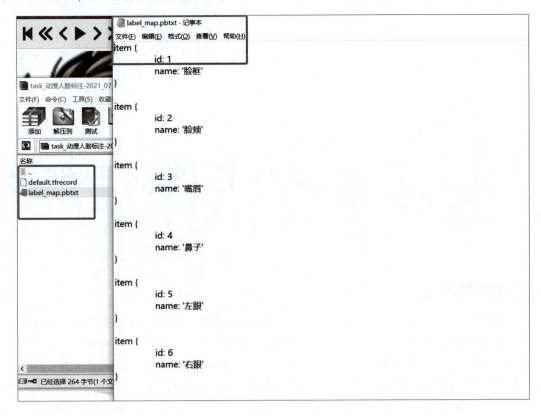

图 2.14　查看标签映射文件

项目小结

祝贺你掌握了矩形框标注和关键点标注的方法。在本项目中所做的工作也是一个典型的人脸识别数据标注过程，人脸识别已经广泛应用在公共安全、电子支付、身份验证等场景。而随着人脸识别算法的不断改进，特征点位也从最初的 4 点、5 点到 21 点、29 点、68 点，以及现在常见的 186 点、270 点等。由于这个项目是卡通图像，因此项目经理要求按照 21 点特征点位标注，数据标注完成后，你的团队还需要做后续的数据训练等步骤，才能实现特征提取的工作。

多学一点：人脸识别的过程中有 4 个关键任务：人脸检测、人脸对齐、人脸编码（即特征提取）、人脸匹配。其中人脸检测的目的是寻找图片中人脸的位置；人脸对齐是将不同角度的人脸图像对齐成同一种标准的形状，比如先定位人脸上的特征点，然后通过几何变换，使各个特征点对齐；接下来就是人脸编码也就是特征提取，人脸图像的像素值会被转换成可判别的特征向量；最后在人脸匹配过程中，两个特征向量会进行比较，从而得到一个相似度分数，该分数给出了两者属于同一个主体的可能性。为了防止欺骗，人脸识别通常还会加入生物识别、活体检测等任务。祝愿你在未来的学习中掌握更多的技能，在实际工作中选择最有效的数据标注方法，以终为始，不断提高标注的技巧，成为一名优秀的工程师。

项目 3

宠物管理猫狗检测

项目背景

学电子信息的你毕业后进入了一家系统集成公司,公司正在从传统的计算机和网络系统集成公司升级转型为一个具备 AI 能力的信息系统集成公司。由于你在学校学习了图像识别相关知识,这下可派上了用场。团队承接了一个系统集成的项目,客户是北京三里屯某宠物店,养了数十只猫和狗,200 m^2 的店面分成三个区域,一个是猫生活区,一个是狗生活区,还有一个面积较大的是客户活动区。客户遇到的问题是,猫和狗会趁人不注意时走错生活区,管理员除了每隔一段时间查看之外,还希望引入一个更自动的方法,让店里的摄像机判断是否有宠物待在了不该待的生活区。承接这个项目后,项目经理组建了团队,要做的第一个工作就是通过图像识别猫和狗,然后判断是不是出现在了不该出现的区域,最后通过与硬件设备的关联实现自动检测和语音报警提示。项目经理分配给你的任务是:根据拿到的数据和预训练模型,生成一个准确率较高的图像识别模型,实现输入一张图片能够识别猫和狗。

提示:猫狗识别对于计算机视觉来说是最典型的任务之一,项目经理已经提供了数百张照片和一组预训练模型。在这个项目中,要亲自动手完成从数据标注到模型训练再到模型导出等任务,实现可以用于边缘设备部署的模型。

任务 1 数据准备

任务描述

在本任务中,将获得项目团队提供的猫狗图片数据集,并把这些图片数据导入人工智能数据处理平台中,分别保存在训练、验证、测试等不同的目录中,为后续数据标注工作做好准备。

项目 3　宠物管理猫狗检测

步骤 1　数据采集

项目经理已经准备好了相关图片数据。数据采集途径有多种,可以下载已有的数据集,也可以使用自己工作中的照片、图片等制作新的数据集。

步骤 2　数据整理

(1) 把数据下载到工作目录,解压缩。
(2) 在终端命令行窗口中执行以下操作。
＊注意第二行命令需要把地址修改成对应的资源平台地址。

```
$ cd ~/data
$ wget http://172.16.33.72/dataset/cat_dog.tar.gz
$ tar zxvf cat_dog.tar.gz
```

(3) 在终端命令行窗口中执行以下操作,查看解压缩后的目录,如图 3.1 所示。

```
$ cd ~/data/cat_dog
$ ls
```

图 3.1　查看解压缩后的目录

任务 2　工程环境准备

任务描述

在本任务中,将创建本项目的开发环境,并进行基础配置,满足数据标注、模型训练、评估和部署的基础条件。

步骤 1　创建工程目录

在开发环境中打开为本项目创建的工程目录,在终端命令行窗口中执行以下操作:

```
$ mkdir ~/projects/unit3
$ mkdir ~/projects/unit3/data
$ cd ~/projects/unit3
```

步骤 2　创建开发环境

在 Python 3.6 中创建名为 unit3 的虚拟环境。

```
$ conda create -n unit3 python=3.6
```

输入 y 继续完成,然后执行以下操作激活开发环境。

```
$ conda activate unit3
```

步骤3 配置GPU环境

安装tensorflow-gpu1.15环境。

```
$ conda install tensorflow-gpu=1.15
```

输入y继续完成GPU环境的配置,如图3.2所示。

图3.2 完成GPU环境的配置

步骤4 配置依赖环境

(1)在开发环境中打开/home/student/projects/unit3目录,创建依赖清单文件requirements.txt。

(2)将以下内容写到requirements.txt清单文件中,然后执行命令,安装依赖库环境,完成依赖环境的配置,如图3.3所示。

图3.3 完成依赖环境的配置

```
# requirements.txt
Cython
contextlib2
matplotlib
pillow
lxml
jupyter
pycocotools
click
PyYAML
joblib
autopep8

#执行以下命令:
$ conda activate unit3
$ pip install -r requirements.txt
```

步骤 5　配置目标检测库环境

（1）安装中育 object_detection 库和中育 slim 库。中育 object_detection 库和中育 slim 库为目标检测模型库，后面的任务将通过这两个库生成基础的目标检测模型。

（2）在终端命令行窗口中执行以下操作，完成后删除安装程序。

＊注意第一行和第三行命令需要把地址修改成对应的资源平台地址。

```
$ wget http://172.16.33.72/dataset/dist/zy_od_1.0.tar.gz
$ pip install zy_od_1.0.tar.gz
$ wget http://172.16.33.72/dataset/dist/zy_slim_1.0.tar.gz
$ pip install zy_slim_1.0.tar.gz
$ rm zy_slim_1.0.tar.gz zy_od_1.0.tar.gz
```

步骤 6　验证环境

在终端命令行窗口中执行以下操作，查看验证结果，如图 3.4 所示。

图 3.4　验证结果

人工智能图像识别项目实践

＊注意需要把地址修改成对应的资源平台地址。

```
$ wget http://172.16.33.72/dataset/script/env_test.py
$ python env_test.py
```

任务 3　猫狗图片数据标注

任务描述

在本任务中，将使用图片标注工具完成数据标注，导出为数据集文件，并保存标签映射文件。

任务操作

步骤 1　添加标注标签

（1）创建名称为"宠物管理猫狗检测"的标注项目，如图 3.5 所示。
（2）添加 2 类标签，分别为 cat、dog。
（3）注意将标签设置为不同的颜色以示区分。

图 3.5　创建标注项目

步骤 2　创建训练集任务

（1）任务名称为"宠物管理猫狗检测训练集"。
（2）任务子集选择 Train。
（3）选择文件使用"连接共享文件"，选中任务 1 中整理的 train 子目录，如图 3.6 所示。
（4）单击"提交"按钮，完成创建。

项目3　宠物管理猫狗检测

图 3.6　创建训练集任务

步骤3　标注训练集数据

（1）打开"宠物管理猫狗检测训练集"，单击左下方的"作业#"进入标注作业，如图 3.7 所示。

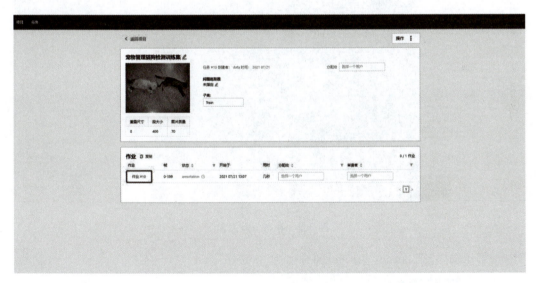

图 3.7　进入训练集标注作业

（2）使用加锁，可以避免对已标注对象的误操作；将一张图片中的对象标注完成后，单击上方工具栏中的"下一帧"按钮继续标注，如图 3.8 所示，直至整个数据集标注完成。

步骤4　导出标注训练集

（1）选择"菜单"→"导出为数据集"→"TFRecord 1.0"命令，如图 3.9 所示。

（2）标注完成的数据导出后是一个压缩包 zip 文件，保存在浏览器默认的下载路径中。将

31

图 3.8 标注图片

图 3.9 导出为数据集

这个文件解压缩,并把 default.tfrecord 重命名为 train.tfrecord。

步骤5 创建验证集任务

(1)任务名称为"宠物管理猫狗检测验证集"。
(2)任务子集选择 Validation。
(3)选择文件使用"连接共享文件",选中任务1中整理的 val 子目录,如图 3.10 所示。
(4)单击"提交"按钮,完成创建。

步骤6 标注验证集数据

(1)打开"宠物管理猫狗检测验证集",单击左下方的"作业"进入标注,如图 3.11 所示。

项目3　宠物管理猫狗检测

（2）将一张图片中的对象标注完成后，单击上方工具栏中的"下一帧"按钮继续标注，直至整个数据集标注完成。

图 3.10　创建验证任务

图 3.11　进入验证集标注作业

步骤 7　导出标注验证集

（1）选择"菜单"→"导出为数据集"→"TFRecord 1.0"命令。

（2）标注完成的数据集导出后是一个压缩文件，保存在浏览器默认的下载路径中。

（3）将该文件解压缩后会得到 default.tfrecord 和 label_map.pbtxt 文件，把 default.tfrecord 重命名为 val.tfrecord。

（4）找到之前保存好的 train.tfrecord 文件，并把 val.tfrecord、train.tfrecord、label_map.pbtxt

三个文件存放到一起备用。

步骤8 上传文件

(1) 打开系统提供的 winSCP 工具,找到之前准备好的 val.tfrecord、train.tfrecord、label_map.pbtxt 文件,把这三个文件上传到数据处理平台中的 home/student/projects/unit3/data 目录下,如图 3.12 所示。

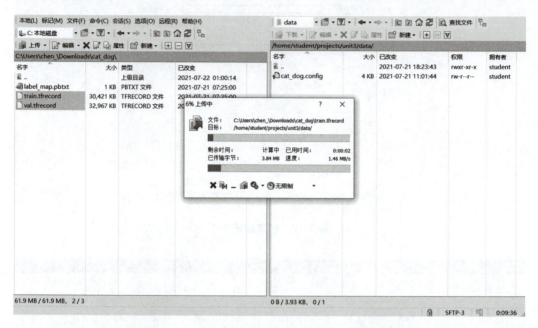

图 3.12 上传标注数据文件

(2) 上传成功后,在平台上可以看到以下三个文件,如图 3.13 所示。

图 3.13 数据标注文件上传成功

任务 4 猫狗检测模型训练

任务描述

在本任务中,将搭建训练模型、配置预训练模型参数,对已标注的数据集进行训练,得到训练模型,并学习使用可视化的工具查看训练结果。

任务操作

步骤 1 搭建模型

(1) 在开发环境中打开准备预训练模型相关目录。

```
$ cd ~/projects/unit3
$ mkdir pretrain_models
$ cd pretrain_models
```

（2）下载算法团队提供的预训练模型，并解压缩。

* 注意需要把地址修改成对应的资源平台地址。

```
$ wget http://172.16.33.72/dataset/dist/zy_ptm_u3.tar.gz
$ tar zxvf zy_ptm_u3.tar.gz
$ rm zy_ptm_u3.tar.gz
```

步骤2　配置训练模型

在开发环境中打开/home/student/projects/unit3/data 目录，创建训练模型配置文件 cat_dog.config。

（1）主干网络配置。主干网络是整个模型训练的基础，标记了当前模型识别的物体类别等重要信息。本项目猫狗识别为两类，因此 num_classes 为 2。

```
num_classes:2
box_coder {
  faster_rcnn_box_coder {
    y_scale:10.0
    x_scale:10.0
    height_scale:5.0
    width_scale:5.0
  }
}
matcher {
  argmax_matcher {
    matched_threshold:0.5
    unmatched_threshold:0.5
    ignore_thresholds:false
    negatives_lower_than_unmatched:true
    force_match_for_each_row:true
  }
}
similarity_calculator {
  iou_similarity {
  }
}
```

（2）先验框配置和图片分辨率配置。image_resizer 表示模型输入图片分辨率，此处为标准的 300×300 像素，因此 height 为 300，width 为 300。

```
anchor_generator {
  ssd_anchor_generator {
    num_layers:6
    min_scale:0.2
    max_scale:0.95
    aspect_ratios:1.0
    aspect_ratios:2.0
```

```
      aspect_ratios:0.5
      aspect_ratios:3.0
      aspect_ratios:0.3333
    }
  }
  image_resizer {
    fixed_shape_resizer {
      height:300
      width:300
    }
  }
```

(3)边界预测框配置。

```
box_predictor {
  convolutional_box_predictor {
    min_depth:0
    max_depth:0
    num_layers_before_predictor:0
    use_dropout:false
    dropout_keep_probability:0.8
    kernel_size:1
    box_code_size:4
    apply_sigmoid_to_scores:false
    conv_hyperparams {
      activation:RELU_6,
      regularizer {
        l2_regularizer {
          weight:0.00004
        }
      }
      initializer {
        truncated_normal_initializer {
          stddev:0.03
          mean:0.0
        }
      }
      batch_norm {
        train:true,
        scale:true,
        center:true,
        decay:0.9997,
        epsilon:0.001,
      }
    }
  }
}
```

(4)特征提取网络配置。

```
feature_extractor {
  type:'ssd_mobilenet_v2'
  min_depth:16
  depth_multiplier:1.0
  conv_hyperparams {
    activation:RELU_6,
    regularizer {
      l2_regularizer {
        weight:0.00004
      }
    }
    initializer {
      truncated_normal_initializer {
        stddev:0.03
        mean:0.0
      }
    }
    batch_norm {
      train:true,
      scale:true,
      center:true,
      decay:0.9997,
      epsilon:0.001,
    }
  }
}
```

(5)模型损失函数配置。

```
loss {
  classification_loss {
    weighted_sigmoid {
    }
  }
  localization_loss {
    weighted_smooth_l1 {
    }
  }
  hard_example_miner {
    num_hard_examples:3000
    iou_threshold:0.99
    loss_type:CLASSIFICATION
    max_negatives_per_positive:3
    min_negatives_per_image:3
  }
  classification_weight:1.0
  localization_weight:1.0
}
```

```
normalize_loss_by_num_matches:true
post_processing {
  batch_non_max_suppression {
    score_threshold:1e-8
    iou_threshold:0.6
    max_detections_per_class:100
    max_total_detections:100
  }
  score_converter:SIGMOID
}
```

（6）训练集数据配置。batch_size 代表批处理每次迭代的数据量，initial_learning_rate 代表初始学习率，fine_tune_checkpoint 指向预训练模型文件，input_path 指向训练集的 tfrecord 文件，label_map_path 指向标签映射文件。

```
train_config:{
  batch_size:12
  optimizer {
    rms_prop_optimizer:{
      learning_rate:{
        exponential_decay_learning_rate {
          initial_learning_rate:0.004
          decay_steps:1000
          decay_factor:0.95
        }
      }
      momentum_optimizer_value:0.9
      decay:0.9
      epsilon:1.0
    }
  }
  fine_tune_checkpoint:"pretrain_models/zy_ptm_u3/model.ckpt"
  fine_tune_checkpoint_type:  "detection"
  num_steps:2000
  data_augmentation_options {
    random_horizontal_flip {
    }
  }
  data_augmentation_options {
    ssd_random_crop {
    }
  }
}

train_input_reader:{
  tf_record_input_reader {
    input_path:"data/train.tfrecord"
  }
  label_map_path:"data/label_map.pbtxt"
}
```

(7)验证集数据配置。num_examples 代表验证集样本数量,input_path 指向验证集的 tfrecord 文件,label_map_path 指向标签映射文件。

```
eval_config:{
  num_examples:50
  max_evals:1
}

eval_input_reader:{
  tf_record_input_reader {
    input_path:"data/val.tfrecord"
  }
  label_map_path:"data/label_map.pbtxt"
  shuffle:false
  num_readers:1
}
```

模型配置文件 cat_dog.config 的完整源代码文件参见 U3-cat_dog.config.pdf。

cat_dog.config

步骤3　创建训练程序

在开发环境中打开/home/student/projects/unit3/目录,创建训练程序 train.py。
(1)导入训练所需模块和函数。

```
import functools
import json
import os
import tensorflow as tf
from object_detection.builders import dataset_builder
from object_detection.builders import graph_rewriter_builder
from object_detection.builders import model_builder
from object_detection.legacy import trainer
from object_detection.utils import config_util
```

(2)定义输入参数。

```
os.environ["TF_CPP_MIN_LOG_LEVEL"] = '3'
tf.logging.set_verbosity(tf.logging.INFO)
flags = tf.app.flags
flags.DEFINE_string('master','','')
flags.DEFINE_integer('task',0,'task id')
flags.DEFINE_integer('num_clones',1,'')
flags.DEFINE_boolean('clone_on_cpu',False,'')
flags.DEFINE_integer('worker_replicas',1,'')
```

```
flags.DEFINE_integer('ps_tasks',0,'')
flags.DEFINE_string('train_dir','','Directory to save the checkpoints and
training summaries.')
flags.DEFINE_string('pipeline_config_path','','Path to a pipeline config.')
flags.DEFINE_string('train_config_path','','Path to a train_pb2.TrainConfig.')
flags.DEFINE_string('input_config_path','','Path to an input_reader_pb2.InputReader.')
flags.DEFINE_string('model_config_path','','Path to a model_pb2.DetectionModel.')
FLAGS = flags.FLAGS
```

(3)训练主函数:加载模型配置。

```
@tf.contrib.framework.deprecated(None,'Use object_detection/model_main.py.')
def main(_):
  assert FLAGS.train_dir,''train_dir' is missing.'
  if FLAGS.task = = 0:tf.gfile.MakeDirs(FLAGS.train_dir)
  if FLAGS.pipeline_config_path:
    configs = config_util.get_configs_from_pipeline_file(FLAGS.pipeline_config_path)
    if FLAGS.task = = 0:
      tf.gfile.Copy(FLAGS.pipeline_config_path,
                    os.path.join(FLAGS.train_dir,'pipeline.config'),
                    overwrite = True)
  else:
    configs = config_util.get_configs_from_multiple_files(
        model_config_path = FLAGS.model_config_path,
        train_config_path = FLAGS.train_config_path,
        train_input_config_path = FLAGS.input_config_path)
    if FLAGS.task = = 0:
      for name,config in [('model.config',FLAGS.model_config_path),
                          ('train.config',FLAGS.train_config_path),
                          ('input.config',FLAGS.input_config_path)]:
        tf.gfile.Copy(config,os.path.join(FLAGS.train_dir,name),overwrite = True)

  model_config = configs['model']
  train_config = configs['train_config']
  input_config = configs['train_input_config']
  model_fn = functools.partial(
      model_builder.build,
      model_config = model_config,
      is_training = True)
```

(4)训练主函数:设计模型线程和迭代循环。

```
def get_next(config):
  return dataset_builder.make_initializable_iterator(
    dataset_builder.build(config)).get_next()

  create_input_dict_fn = functools.partial(get_next,input_config)
  env = json.loads(os.environ.get('TF_CONFIG','{}'))
  cluster_data = env.get('cluster',None)
  cluster = tf.train.ClusterSpec(cluster_data) if cluster_data else None
```

```
task_data = env.get('task',None) or {'type':'master','index':0}
task_info = type('TaskSpec',(object,),task_data)

ps_tasks = 0
worker_replicas = 1
worker_job_name = 'lonely_worker'
task = 0
is_chief = True
master = ''

if cluster_data and 'worker' in cluster_data:
    worker_replicas = len(cluster_data['worker']) + 1
if cluster_data and 'ps' in cluster_data:
    ps_tasks = len(cluster_data['ps'])

if worker_replicas > 1 and ps_tasks < 1:
    raise ValueError('At least 1 ps task is needed for distributed training.')

if worker_replicas >= 1 and ps_tasks > 0:
    server = tf.train.Server(tf.train.ClusterSpec(cluster),protocol = 'grpc',
                             job_name = task_info.type,
                             task_index = task_info.index)
    if task_info.type == 'ps':
        server.join()
        return
    worker_job_name = '% s/task:% d' % (task_info.type,task_info.index)
    task = task_info.index
    is_chief = (task_info.type == 'master')
    master = server.target
```

（5）训练主函数：记录训练日志，配置训练函数参数。

```
graph_rewriter_fn = None
    if 'graph_rewriter_config' in configs:
        graph_rewriter_fn = graph_rewriter_builder.build(
            configs['graph_rewriter_config'],is_training = True)

    trainer.train(
        create_input_dict_fn,
        model_fn,
        train_config,
        master,
        task,
        FLAGS.num_clones,
        worker_replicas,
        FLAGS.clone_on_cpu,
        ps_tasks,
        worker_job_name,
        is_chief,
```

```
        FLAGS.train_dir,
        graph_hook_fn=graph_rewriter_fn)
    print("模型训练完成!")

if __name__ == '__main__':
    tf.app.run()
```

训练程序 train.py 的完整源代码文件参见 U3-train.py.pdf。

完整源代码

train.py

步骤4 训练模型

运行训练程序 train.py,如图 3.14 所示,程序读取配置文件 cat_dog.config 中定义的训练模型、训练参数、数据集,把训练日志和检查点保存到 checkpoint 目录中。

```
$ conda activate unit3
$ python train.py --logtostderr --train_dir checkpoint --pipeline_config_path data/cat_dog.config
```

图 3.14 训练模型

步骤5 可视化训练过程

在训练过程中打开 tensorboard 可以查看训练日志,如图 3.15 所示。训练日志中记录了模型分类损失、回归损失和总损失量的变化,通过 Losses 选项中的图表可以看到训练过程中的损

项目3　宠物管理猫狗检测

失在不断变化,越到后面损失越小,说明模型对训练数据的拟合度越来越高。

＊注意需要把地址修改成对应的数据处理服务器地址,然后在浏览器中输入对应地址和端口号查看。

```
$ tensorboard --host 172.16.33.11 --port 8889 --logdir checkpoint/
```

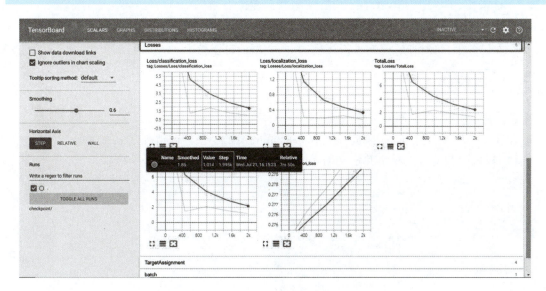

图 3.15　查看训练日志

步骤6　查看训练结果

进入 checkpoint 子目录,可以看到生成了多组模型文件,如图 3.16 所示。

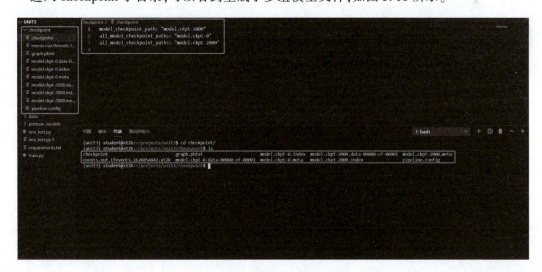

图 3.16　生成多组模型文件

其中:
- model. ckpt-xxxx. meta 文件保存了计算图也就是神经网络的结构;
- model. ckpt-xxxx. data-xxxx 文件保存了模型的变量;

- model.ckpt-xxxx.index 文件保存了神经网络索引映射文件。

任务5　猫狗检测模型评估

任务描述

在本任务中,将对训练模型进行评估,判断模型的可用性。

任务操作

步骤1　创建评估程序

在开发环境中打开/home/student/projects/unit3/目录,创建评估程序 eval.py。

(1) 导入模型的各个模块并定义输入参数。

```
import functools
import os
import tensorflow as tf
from object_detection.builders import dataset_builder
from object_detection.builders import graph_rewriter_builder
from object_detection.builders import model_builder
from object_detection.legacy import evaluator
from object_detection.utils import config_util
from object_detection.utils import label_map_util

os.environ["TF_CPP_MIN_LOG_LEVEL"] = '3'
tf.compat.v1.logging.set_verbosity(tf.compat.v1.logging.ERROR)
flags = tf.app.flags
flags.DEFINE_boolean('eval_training_data',False,'')
flags.DEFINE_string('checkpoint_dir','','')
flags.DEFINE_string('eval_dir','','Directory to write eval summaries.')
flags.DEFINE_string('pipeline_config_path','','Path to a pipeline config.')
flags.DEFINE_string('eval_config_path','','')
flags.DEFINE_string('input_config_path','','')
flags.DEFINE_string('model_config_path','','')
flags.DEFINE_boolean('run_once',False,'')
FLAGS = flags.FLAGS
```

(2) 评估主函数:加载模型配置。

```
@tf.contrib.framework.deprecated(None,'Use object_detection/model_main.py.')
def main(unused_argv):
    assert FLAGS.checkpoint_dir,'checkpoint_dir'is missing.'
    assert FLAGS.eval_dir,''eval_dir' is missing.'
    tf.gfile.MakeDirs(FLAGS.eval_dir)
    if FLAGS.pipeline_config_path:
        configs = config_util.get_configs_from_pipeline_file(
            FLAGS.pipeline_config_path)
        tf.gfile.Copy(
```

```
            FLAGS.pipeline_config_path,
            os.path.join(FLAGS.eval_dir,'pipeline.config'),
            overwrite=True)
    else:
        configs=config_util.get_configs_from_multiple_files(
            model_config_path=FLAGS.model_config_path,
            eval_config_path=FLAGS.eval_config_path,
            eval_input_config_path=FLAGS.input_config_path)
        for name,config in [('model.config',FLAGS.model_config_path),
                            ('eval.config',FLAGS.eval_config_path),
                            ('input.config',FLAGS.input_config_path)]:
            tf.gfile.Copy(config,os.path.join(FLAGS.eval_dir,name),overwrite=True)

    model_config=configs['model']
    eval_config=configs['eval_config']
    input_config=configs['eval_input_config']
    if FLAGS.eval_training_data:
        input_config=configs['train_input_config']

    model_fn=functools.partial(
        model_builder.build,model_config=model_config,is_training=False)
```

(3) 评估主函数：定义评估循环，并记录评估日志。

```
    def get_next(config):
        return dataset_builder.make_initializable_iterator(
            dataset_builder.build(config)).get_next()

    create_input_dict_fn=functools.partial(get_next,input_config)

    categories=label_map_util.create_categories_from_labelmap(
        input_config.label_map_path)

    if FLAGS.run_once:
        eval_config.max_evals=1

    graph_rewriter_fn=None
    if 'graph_rewriter_config' in configs:
        graph_rewriter_fn=graph_rewriter_builder.build(
            configs['graph_rewriter_config'],is_training=False)
```

(4) 评估主函数：配置评估函数参数。

```
    evaluator.evaluate(
        create_input_dict_fn,
        model_fn,
        eval_config,
        categories,
        FLAGS.checkpoint_dir,
        FLAGS.eval_dir,
```

```
        graph_hook_fn = graph_rewriter_fn)
    print("模型评估完成!")

if __name__ == '__main__':
    tf.app.run()
```

评估程序 eval.py 的完整源代码文件参见 U3-eval.py.pdf。

eval.py

步骤 2　评估模型

运行评估程序 eval.py，如图 3.17 所示，程序读取配置文件 cat_dog.config 中定义的训练模型、训练参数、数据集，读取 checkpoint 目录中的训练结果，把评估结果保存到 evaluation 目录中。

```
$ conda activate unit3
$ python eval.py --logtostderr --checkpoint_dir checkpoint --eval_dir evaluation --pipeline_config_path data/cat_dog.config
```

图 3.17　评估模型

在评估过程中，可以看到对不同类别的评估结果。

步骤3 查看评估结果

利用 tensorboard 工具查看评估结果,如图 3.18 所示。

* 注意需要把地址修改成对应的数据处理服务器地址,然后在浏览器中输入对应地址和端口号查看。

```
$ tensorboard --host 172.16.33.11 --port 8889 --logdir evaluation/
```

步骤4 分析模型可用性

在浏览器中查看各类别的平均精确度(AP)值,越接近1说明模型的可用性越高。此时图上显示,step 是 2k,说明这个模型是训练到 2 000 步时保存下来的,对应 model.ckpt-2000 训练模型。

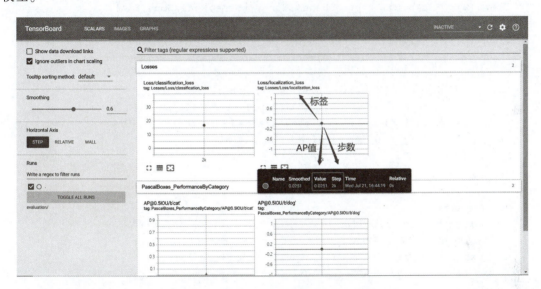

图 3.18 查看评估结果

任务6　猫狗检测模型测试

 任务描述

在本任务中,将把已经评估为可用性较强的模型,导出成为可测试的冻结图模型,用测试数据进行测试。

任务操作

步骤1 创建导出程序

在开发环境中打开/home/student/projects/unit3/目录,创建导出程序 export_fz.py。

(1)导入模型转换模块,定义输入参数。

```python
import os
import tensorflow as tf
from google.protobuf import text_format
from object_detection import exporter
from object_detection.protos import pipeline_pb2

os.environ["TF_CPP_MIN_LOG_LEVEL"] = '3'
tf.compat.v1.logging.set_verbosity(tf.compat.v1.logging.ERROR)
slim = tf.contrib.slim
flags = tf.app.flags

flags.DEFINE_string('input_type','image_tensor','')
flags.DEFINE_string('input_shape',None,'[None,None,None,3]')
flags.DEFINE_string('pipeline_config_path',None,'Path to a pipeline config.')
flags.DEFINE_string('trained_checkpoint_prefix',None,'path/to/model.ckpt')
flags.DEFINE_string('output_directory',None,'Path to write outputs.')
flags.DEFINE_string('config_override','','')
flags.DEFINE_boolean('write_inference_graph',False,'')
tf.app.flags.mark_flag_as_required('pipeline_config_path')
tf.app.flags.mark_flag_as_required('trained_checkpoint_prefix')
tf.app.flags.mark_flag_as_required('output_directory')
FLAGS = flags.FLAGS
```

(2)转换模型主函数,调用模型转换函数。

```python
def main(_):
    pipeline_config = pipeline_pb2.TrainEvalPipelineConfig()
    with tf.gfile.GFile(FLAGS.pipeline_config_path,'r') as f:
        text_format.Merge(f.read(),pipeline_config)
    text_format.Merge(FLAGS.config_override,pipeline_config)
    if FLAGS.input_shape:
        input_shape = [
            int(dim) if dim != '-1' else None
            for dim in FLAGS.input_shape.split(',')
        ]
    else:
        input_shape = None
    exporter.export_inference_graph(
        FLAGS.input_type,pipeline_config,FLAGS.trained_checkpoint_prefix,
        FLAGS.output_directory,input_shape=input_shape,
        write_inference_graph=FLAGS.write_inference_graph)
    print("模型转换完成!")

if __name__ == '__main__':
    tf.app.run()
```

导出程序 export_fz.py 的完整源代码文件参见 U3-export_fz.py.pdf。

完整源代码

export_fz.py

步骤2 导出冻结图模型

运行导出程序 export_fz.py，如图 3.19 所示，程序将读取配置文件 cat_dog.config 中定义的配置，读取 checkpoint 目录中的 model.ckpt-2000 训练模型，导出为冻结图模型，并保存到 frozen_models 目录中。

```
$ conda activate unit3
$ python export_fz.py --input_type image_tensor --pipeline_config_path data/cat_dog.config --trained_checkpoint_prefix checkpoint/model.ckpt-2000 --output_directory frozen_models
```

图 3.19 导出冻结图模型

步骤3 创建测试程序

在开发环境中打开/home/student/projects/unit3/目录，创建测试文件 detect.py。
(1) 导入测试所需模块和可视化函数，定义输入参数。

```
import numpy as np
import os
import tensorflow as tf
```

```python
import matplotlib.pyplot as plt
from PIL import Image
from object_detection.utils import label_map_util
from object_detection.utils import visualization_utils as vis_util
from object_detection.utils import ops as utils_ops

os.environ["TF_CPP_MIN_LOG_LEVEL"] = '3'
tf.compat.v1.logging.set_verbosity(tf.compat.v1.logging.ERROR)
detect_img = '/home/student/data/cat_dog/test/000215.jpg'
result_img = '/home/student/projects/unit3/img/000215_result.jpg'
MODEL_NAME = 'frozen_models'
PATH_TO_FROZEN_GRAPH = MODEL_NAME + '/frozen_inference_graph.pb'
PATH_TO_LABELS = 'data/label_map.pbtxt'
```

（2）加载模型计算图和数据标签。

```python
detection_graph = tf.Graph()
with detection_graph.as_default():
    od_graph_def = tf.compat.v1.GraphDef()
    with tf.io.gfile.GFile(PATH_TO_FROZEN_GRAPH,'rb') as fid:
        serialized_graph = fid.read()
        od_graph_def.ParseFromString(serialized_graph)
        tf.import_graph_def(od_graph_def,name = '')
category_index = label_map_util.create_category_index_from_labelmap
        (PATH_TO_LABELS,use_display_name = True)
```

（3）图片数据转换函数。

```python
def load_image_into_numpy_array(image):
    (im_width,im_height) = image.size
    return np.array(image.getdata()).reshape((im_height,im_width,3)).astype(np.uint8)
```

（4）单张图像检测函数。

```python
def run_inference_for_single_image(image,graph):
    with graph.as_default():
        with tf.compat.v1.Session() as sess:
            ops = tf.compat.v1.get_default_graph().get_operations()
            all_tensor_names = {output.name for op in ops for output in op.outputs}
            tensor_dict = {}
            for key in ['num_detections','detection_boxes','detection_scores',
                'detection_classes','detection_masks']:
                tensor_name = key + ':0'
                if tensor_name in all_tensor_names:
                    tensor_dict[key] = tf.compat.v1.get_default_graph().get_tensor_by_name(tensor_name)
            if 'detection_masks' in tensor_dict:
                detection_boxes = tf.squeeze(tensor_dict['detection_boxes'],[0])
                detection_masks = tf.squeeze(tensor_dict['detection_masks'],[0])
                real_num_detection = tf.cast(tensor_dict['num_detections'][0],tf.int32)
                detection_boxes = tf.slice(detection_boxes,[0,0],[real_num_detection,-1])
```

```python
            detection_masks = tf.slice(detection_masks,[0,0,0],[real_num_detection,-1,-1])
            detection_masks_reframed = utils_ops.reframe_box_masks_to_image_masks(
                detection_masks,detection_boxes,image.shape[1],image.shape[2])
            detection_masks_reframed = tf.cast(tf.greater(detection_masks_reframed,0.5),tf.uint8)
            tensor_dict['detection_masks'] = tf.expand_dims(detection_masks_reframed,0)
        image_tensor = tf.compat.v1.get_default_graph().get_tensor_by_name('image_tensor:0')

        output_dict = sess.run(tensor_dict,feed_dict = {image_tensor:image})
        output_dict['num_detections'] = int(output_dict['num_detections'][0])
        output_dict['detection_classes'] = output_dict['detection_classes'][0].astype(np.int64)
        output_dict['detection_boxes'] = output_dict['detection_boxes'][0]
        output_dict['detection_scores'] = output_dict['detection_scores'][0]
        if 'detection_masks' in output_dict:
            output_dict['detection_masks'] = output_dict['detection_masks'][0]
    return output_dict
```

(5)输入图片数据,检测输入数据,保存检测结果图。

```python
image = Image.open(detect_img)
image_np = load_image_into_numpy_array(image)
# 转化输入图片为 shape = [1,None,None,3]
image_np_expanded = np.expand_dims(image_np,axis = 0)
output_dict = run_inference_for_single_image(image_np_expanded,detection_graph)
vis_util.visualize_boxes_and_labels_on_image_array(
    image_np,
    output_dict['detection_boxes'],
    output_dict['detection_classes'],
    output_dict['detection_scores'],
    category_index,
    instance_masks = output_dict.get('detection_masks'),
    use_normalized_coordinates = True,
    line_thickness = 6)
plt.figure()
plt.axis('off')
plt.imshow(image_np)
plt.savefig(result_img,bbox_inches = 'tight',pad_inches = 0)
print("测试%s 完成,结果保存在%s" % (detect_img,result_img))
```

测试程序 detect.py 的完整源代码文件参见 U3-detect.py.pdf。

完整源代码

detect.py

步骤4 测试并查看结果

创建 img 目录存放测试结果,运行测试程序 detect.py,如图 3.20 所示,并查看输出的结果图片,如图 3.21 所示。

```
$ mkdir img
$ python detect.py
```

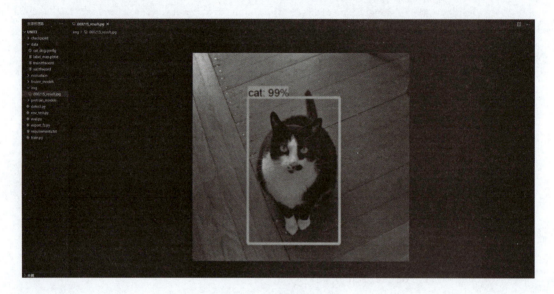

图 3.20 测试模型

图 3.21 查看测试结果

任务7 猫狗检测模型部署

任务描述

在本任务中,将把经过测试确认可用的模型,转换成为应用部署支持的格式,然后将模型文件部署到边缘计算设备上,并实现人工智能应用的集成。

任务操作

步骤1 创建导出程序

在开发环境中打开 /home/student/projects/unit3/ 目录,创建导出程序 export_pb.py。
(1) 导入模型转换模块,定义输入参数。

```python
import os
import tensorflow as tf
from google.protobuf import text_format
from object_detection import export_tflite_ssd_graph_lib
from object_detection.protos import pipeline_pb2

os.environ["TF_CPP_MIN_LOG_LEVEL"] = '3'
tf.compat.v1.logging.set_verbosity(tf.compat.v1.logging.ERROR)
flags = tf.app.flags
flags.DEFINE_string('output_directory', None, 'Path to write outputs. ')
flags.DEFINE_string('pipeline_config_path', None, '')
flags.DEFINE_string('trained_checkpoint_prefix', None, 'Checkpoint prefix. ')
flags.DEFINE_integer('max_detections', 10, '')
flags.DEFINE_integer('max_classes_per_detection', 1, '')
flags.DEFINE_integer('detections_per_class', 100, '')
flags.DEFINE_bool('add_postprocessing_op', True, '')
flags.DEFINE_bool('use_regular_nms', False, '')
flags.DEFINE_string('config_override', '', '')
FLAGS = flags.FLAGS
```

（2）调用模型转换函数，完成模型转换。

```python
def main(argv):
  flags.mark_flag_as_required('output_directory')
  flags.mark_flag_as_required('pipeline_config_path')
  flags.mark_flag_as_required('trained_checkpoint_prefix')

  pipeline_config = pipeline_pb2.TrainEvalPipelineConfig()

  with tf.gfile.GFile(FLAGS.pipeline_config_path, 'r') as f:
    text_format.Merge(f.read(), pipeline_config)
  text_format.Merge(FLAGS.config_override, pipeline_config)
  export_tflite_ssd_graph_lib.export_tflite_graph(
      pipeline_config, FLAGS.trained_checkpoint_prefix, FLAGS.output_directory,
      FLAGS.add_postprocessing_op, FLAGS.max_detections,
      FLAGS.max_classes_per_detection, FLAGS.use_regular_nms)
  print("模型转换完成!")

if __name__ == '__main__':
  tf.app.run(main)
```

导出程序 export_pb.py 的完整源代码文件参见 U3-export_pb.py.pdf。

完整源代码

export_pb.py

步骤2　导出pb文件

运行导出程序 export_pb.py，读取配置文件 cat_dog.config 中定义的参数，读取 checkpoint 目录中的训练结果，把 tflite_pb 模型图保存到 tflite_models 目录中。

```
$ conda activate unit3
$ python export_pb.py --pipeline_config_path data/cat_dog.config --trained_checkpoint_prefix checkpoint/model.ckpt-2000 --output_directory tflite_models
```

步骤3　创建转换程序

在开发环境中打开 /home/student/projects/unit3/ 目录，创建转换程序 pb_to_tflite.py。

(1) 导入模块，定义输入参数。

```python
import os
import tensorflow as tf

os.environ["TF_CPP_MIN_LOG_LEVEL"] = '3'
tf.compat.v1.logging.set_verbosity(tf.compat.v1.logging.ERROR)
flags = tf.app.flags
flags.DEFINE_string('pb_path','tflite_models/tflite_graph.pb','tflite pb file.')
flags.DEFINE_string('tflite_path','tflite_models/zy_ssd.tflite','output tflite.')
FLAGS = flags.FLAGS
```

(2) 转换为 tflite 模型。

```python
def convert_pb_to_tflite(pb_path,tflite_path):
    # 模型输入节点
    input_tensor_name = ["normalized_input_image_tensor"]
    input_tensor_shape = {"normalized_input_image_tensor":[1,300,300,3]}
    # 模型输出节点
    classes_tensor_name =['TFLite_Detection_PostProcess','TFLite_Detection_PostProcess:1',
            'TFLite_Detection_PostProcess:2','TFLite_Detection_PostProcess:3']
    # 转换为 tflite 模型
    converter = tf.lite.TFLiteConverter.from_frozen_graph(pb_path,
                                        input_tensor_name,
                                        classes_tensor_name,
                                        input_tensor_shape)

    converter.allow_custom_ops = True
    converter.optimizations = [tf.lite.Optimize.DEFAULT]
    tflite_model = converter.convert()
```

(3) tflite 模型写入。

```python
    converter.allow_custom_ops = True
    converter.optimizations = [tf.lite.Optimize.DEFAULT]
    tflite_model = converter.convert()
    # 模型写入
    if not tf.gfile.Exists(os.path.dirname(tflite_path)):
        tf.gfile.MakeDirs(os.path.dirname(tflite_path))
    with open(tflite_path,"wb") as f:
```

```
        f.write(tflite_model)
    print("Save tflite model at % s" % tflite_path)
    print("模型转换完成!")

if __name__ = = '__main__':
    convert_pb_to_tflite(FLAGS.pb_path,FLAGS.tflite_path)
```

转换程序 pb_to_tflite.py 的完整源代码文件参见 U3 – pb_to_tflite.py.pdf。

完整源代码

pb_to_tflite.py

步骤 4　转换 tflite 文件

运行程序 pb_to_tflite.py。

```
$ python pb_to_tflite.py
```

步骤 5　创建推理执行程序

在开发环境中打开/home/student/projects/unit3/tflite_models 目录，创建推理执行程序 func_detection_img.py。

（1）导入所用模块。

```
import os
import cv2
import numpy as np
import sys
import glob
import importlib.util
import base64
```

（2）定义模型和数据推理器。

```
def update_image(image_data,GRAPH_NAME = 'zy_ssd.tflite',min_conf_threshold = 0.5,
        use_TPU = False,model_dir = 'util'):
    from tflite_runtime.interpreter import Interpreter
    CWD_PATH = os.getcwd()
    PATH_TO_CKPT = os.path.join(CWD_PATH,model_dir,GRAPH_NAME)

    labels = ['cat','dog']

    interpreter = Interpreter(model_path = PATH_TO_CKPT)

    interpreter.allocate_tensors()

    input_details = interpreter.get_input_details()
```

```python
    output_details = interpreter.get_output_details()
    height = input_details[0]['shape'][1]
    width = input_details[0]['shape'][2]

    floating_model = (input_details[0]['dtype'] == np.float32)

    input_mean = 127.5
    input_std = 127.5
```

(3)输入图像并转换图像数据为张量。

```python
    # base64 解码
    img_data = base64.b64decode(image_data)
    # 转换为 np 数组
    img_array = np.fromstring(img_data, np.uint8)
    # 转换成 opencv 可用格式
    image = cv2.imdecode(img_array, cv2.COLOR_RGB2BGR)

    image_rgb = cv2.cvtColor(image, cv2.COLOR_BGR2RGB)
    imH, imW, _ = image.shape
    image_resized = cv2.resize(image_rgb, (width, height))
    input_data = np.expand_dims(image_resized, axis=0)

    if floating_model:
        input_data = (np.float32(input_data) - input_mean) / input_std

    interpreter.set_tensor(input_details[0]['index'], input_data)
    interpreter.invoke()

    boxes = interpreter.get_tensor(output_details[0]['index'])[0]
    classes = interpreter.get_tensor(output_details[1]['index'])[0]
    scores = interpreter.get_tensor(output_details[2]['index'])[0]
```

(4)检测图片,并可视化输出结果。

```python
for i in range(len(scores)):
  if((scores[i] > min_conf_threshold) and (scores[i] <= 1.0)):
    ymin = int(max(1, (boxes[i][0] * imH)))
    xmin = int(max(1, (boxes[i][1] * imW)))
    ymax = int(min(imH, (boxes[i][2] * imH)))
    xmax = int(min(imW, (boxes[i][3] * imW)))

    cv2.rectangle(image, (xmin, ymin), (xmax, ymax), (10, 255, 0), 2)
    object_name = labels[int(classes[i])]
    label = '%s:%d%%' % (object_name, int(scores[i] * 100))
    labelSize, baseLine = cv2.getTextSize(label, cv2.FONT_HERSHEY_SIMPLEX, 0.7, 2)
```

```
        label_ymin = max(ymin,labelSize[1] +10)
        cv2.rectangle(image,(xmin,label_ymin-labelSize[1]-10),(xmin + labelSize[0],
            label_ymin + baseLine-10),(255,255,255),cv2.FILLED)
        cv2.putText(image,label,(xmin,label_ymin-7),cv2.FONT_HERSHEY_SIMPLEX,0.7,
(0,0,0),2)

    image_bytes = cv2.imencode('.jpg',image)[1].tostring()
    image_base64 = base64.b64encode(image_bytes).decode()
    return image_base64
```

推理执行程序 func_detection_img.py 的完整源代码文件参见 U3-func_detection_img.py.pdf。

完整源代码

func_detection_img.py

步骤6 部署到边缘设备

把模型 zy_ssd.tflite 文件、推理执行程序 func_detection_img.py 文件复制到边缘计算设备中。

*注意把 IP 地址修改成对应的推理机地址。

```
$ scp tflite_models/zy_ssd.tflite student@172.16.33.118:/home/student/zy-panel-check/util/
$ scp tflite_models/func_detection_img.py student@172.16.33.118:/home/student/zy-panel-check/util/
```

通过平台上的"应用部署"按钮，上传或输入图片 URL 检测，应用部署成功结果如图 3.22 所示。

图 3.22 应用部署成功

项目小结

通过亲自动手完成数据标注、模型训练、模型导出等任务,你实现了一个猫狗识别模型,并部署到边缘计算设备上,测试准确率达到了99%,识别速度为1~2 s,完全可以应用在宠物管理项目上。祝贺你和团队为后续全面完成宠物店的需求提供了可靠的基础。

多学一点:模型优化意义重大。模型越大,计算成本就越高,因此,边缘计算设备的内存、计算能力和耗电量都可能会因此受到限制。所以需要对模型进行优化,使其在边缘设备上能够顺利运行;通过减小模型的大小,减少需要执行的操作数量,从而减少边缘设备计算量;较小的模型也容易转化为更少的内存使用,也就更节能。常用的模型优化技术有剪枝、量化等,量化可以分成量化感知训练和训练后量化,其中训练后量化又包括了训练后动态量化和训练后静态量化。这些优化方法最终并不是为了训练模型,而是为了进行推理。祝愿你在未来的学习中掌握更多的技能,在实际工作中灵活运用,成为一名优秀的工程师。

项目 4

自动驾驶行人检测

项目背景

学电子信息的你进入了毕业季,经过多轮面试,终于获得了一封来自心仪企业的入职通知信。这是一家设计、生产、制造传感器产品的高新企业,国家相关 11 个部门联合印发《智能汽车创新发展战略》后,公司生产的用于车辆自动驾驶的传感器成为大受欢迎的产品。通过对公司的学习和了解,你知道了"行人作为道路交通的主要参与者,为了有效地保护行人安全、及时告警驾驶人,需要采用一定的方法对前方行人进行有效检测和行为预判。这是实现真正自动驾驶的关键要素所在"。还有一周就要入职了,你想利用这段时间实现一个行人检测的原型产品,用来鼓励自己投入这项伟大的事业。通过继续研究,你了解到行人检测技术主要采用如下方法:一是普通前视摄像头传感器或环视摄像头,包括立体视觉和单目视觉;二是红外传感器探测,多用于夜间探测的夜视系统;三是雷达传感器探测,包括前中距离雷达和角雷达;四是多传感信息融合技术。图像识别是你学生阶段最喜欢的课程之一,所以你选择第一个方法来动手实现原型。由于在学校学习过数据采集、数据标注、模型训练、模型导出等基础知识,现在是时候来一场综合能力的实践了。

提示:自动驾驶行人检测是一个典型的目标检测任务,在计算机视觉领域,目标检测任务是找出图像中所感兴趣的目标,确定它们的位置和大小。在这个项目中,要完成的是通过数据标注、数据训练、模型导出等任务,让机器找到未知图片中的人以及他们的位置。

任务 1　数据准备

任务描述

在本任务中,我们将获得项目团队提供的行人图片数据集,并把这些图片数据导入人工智能数据处理平台中,分别保存在训练、验证、测试等不同的目录中,为后续数据标注工作做好准备。

步骤1　数据采集

项目经理已经准备好了相关图片数据。数据采集途径有多种,可以直接下载已有的数据集,也可以使用自己工作中的照片、图片等制作新的数据集。

步骤2　数据整理

把数据下载到工作目录,解压缩。在终端命令行窗口中执行以下操作。

*注意第二行命令需要把地址修改成对应的资源平台地址。

```
$ cd ~/data
$ wget http://172.16.33.72/dataset/person.tar.gz
$ tar zxvf person.tar.gz
```

在终端命令行窗口中执行以下操作,查看解压缩后的目录,如图4.1所示。

```
$ cd ~/data/person
$ ls
```

```
student@xt2k:~/data$ cd person/
student@xt2k:~/data/person$ ls
test  train  val
student@xt2k:~/data/person$
```

图4.1　查看解压缩后的目录

任务2　工程环境准备

任务描述

在本任务中,将创建本项目的开发环境,并做基础配置,满足数据标注、模型训练、评估和部署的基础条件。

步骤1　创建工程目录

在开发环境中打开为本项目创建的工程目录,在终端命令行窗口中执行以下操作:

```
$ mkdir ~/projects/unit4
$ mkdir ~/projects/unit4/data
$ cd ~/projects/unit4
```

步骤2　创建开发环境

(1)在Python 3.6中创建名为unit4的虚拟环境。

```
$ conda create -n unit4 python=3.6
```

(2)输入y继续完成,然后执行以下操作激活开发环境。

```
$ conda activate unit4
```

步骤 3　配置 GPU 环境

(1) 安装 tensorflow-gpu 1.15 环境。

```
$ conda install tensorflow-gpu=1.15
```

(2) 输入 y 继续完成 GPU 环境的配置,如图 4.2 所示。

图 4.2　完成 GPU 环境的配置

步骤 4　配置依赖环境

(1) 在开发环境中打开/home/student/projects/unit4 目录,创建依赖清单文件 requirements.txt。

(2) 把以下内容写到 requirements.txt 清单文件中,然后执行命令,安装依赖库环境,完成依赖环境的配置,如图 4.3 所示。

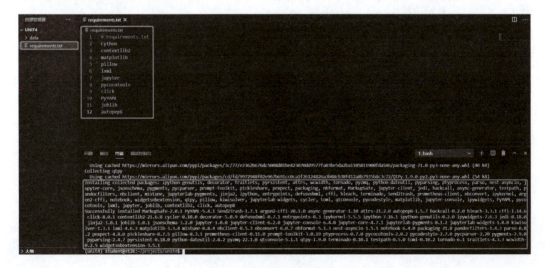

图 4.3　完成依赖环境的配置

```
# requirements.txt
Cython
contextlib2
matplotlib
pillow
lxml
jupyter
pycocotools
click
PyYAML
joblib
autopep8

#执行以下命令：
$ conda activate unit4
$ pip install-r requirements.txt
```

步骤5　配置图像识别库环境

（1）安装中育 object_detection 库和中育 slim 库。中育 object_detection 库和中育 slim 库为目标检测模型库，后面的任务将通过这两个库生成基础的目标检测模型。

（2）在终端命令行窗口中执行以下操作，完成后删除安装程序。

＊注意第一行和第三行命令需要把地址修改成对应的资源平台地址。

```
$ wget http://172.16.33.72/dataset/dist/zy_od_1.0.tar.gz
$ pip install zy_od_1.0.tar.gz
$ wget http://172.16.33.72/dataset/dist/zy_slim_1.0.tar.gz
$ pip install zy_slim_1.0.tar.gz
$ rm zy_od_1.0.tar.gz zy_slim_1.0.tar.gz
```

步骤6　验证环境

在终端命令行窗口中执行以下操作，查看验证结果，如图4.4所示。

＊注意需要把地址修改成对应的资源平台地址。

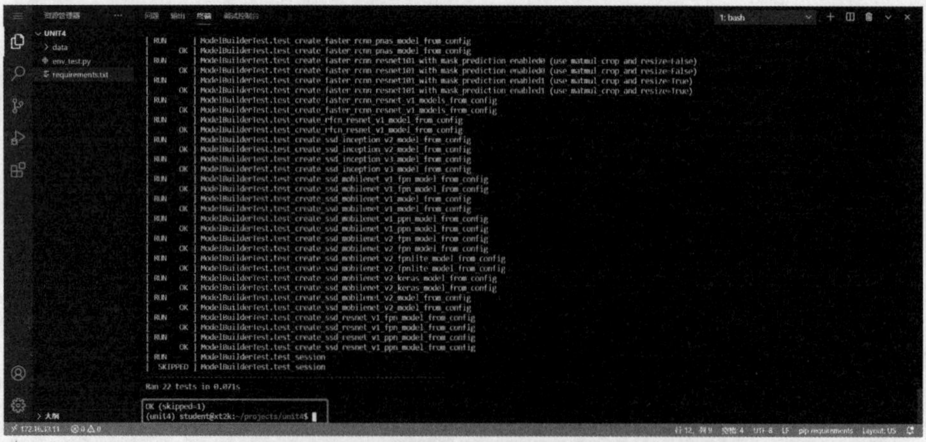

图4.4　验证结果

项目 4 自动驾驶行人检测

```
$ wget http://172.16.33.72/dataset/script/env_test.py
$ python env_test.py
```

任务 3　行人图片数据标注

任务描述

在本任务中，将使用图片标注工具完成数据标注，导出为数据集文件，并保存标签映射文件。

任务操作

步骤 1　添加标注标签

（1）创建名称为"自动驾驶行人检测"的标注项目，如图 4.5 所示。
（2）添加 1 类标签，为 person。
（3）注意设置为不同颜色的标签以示区分。

图 4.5　创建标注项目

步骤 2　创建训练集任务

（1）任务名称为"自动驾驶行人检测训练集"。
（2）任务子集选择 Train。
（3）选择文件使用"连接共享文件"，选中本项目任务 1 中整理的 train 子目录，如图 4.6 所示。
（4）单击"提交"按钮，完成创建。

步骤 3　标注训练集数据

（1）打开"自动驾驶行人检测训练集"，单击左下方的"作业#"进入标注作业，如图 4.7 所示。

63

图 4.6 创建训练任务

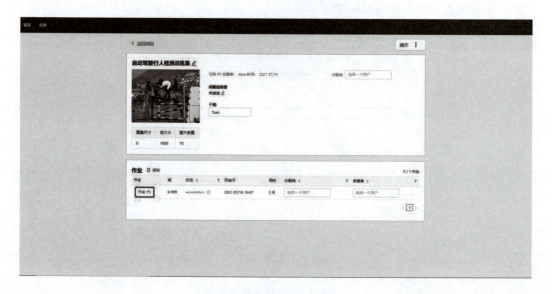

图 4.7 进入训练集标注作业

(2)使用加锁,可以避免对已标注对象的误操作;将一张图片中的对象标注完成后,单击上方工具栏中的"下一帧"按钮继续标注,如图 4.8 所示。

(3)继续标注,直至整个数据集标注完成。

步骤 4 导出标注训练集

(1)选择"菜单"→"导出为数据集"→"TFRecord 1.0"命令,如图 4.9 所示。

(2)标注完成的数据导出后是一个压缩包 zip 文件,保存在浏览器默认的下载路径中。将该文件解压缩,并把 default.tfrecord 重命名为 train.tfrecord。

项目 4　自动驾驶行人检测

图 4.8　标注图片

图 4.9　导出为数据集

步骤 5　创建验证集任务

(1) 任务名称为"自动驾驶行人检测验证集"。
(2) 任务子集选择 Validation。
(3) 选择文件使用"连接共享文件",选中本项目任务 1 中整理的 val 子目录,如图 4.10 所示。
(4) 单击"提交"按钮,完成创建。

步骤 6　标注验证集数据

(1) 打开"自动驾驶行人检测验证集",单击左下方的"作业#"进入标注作业,如图 4.11 所示。

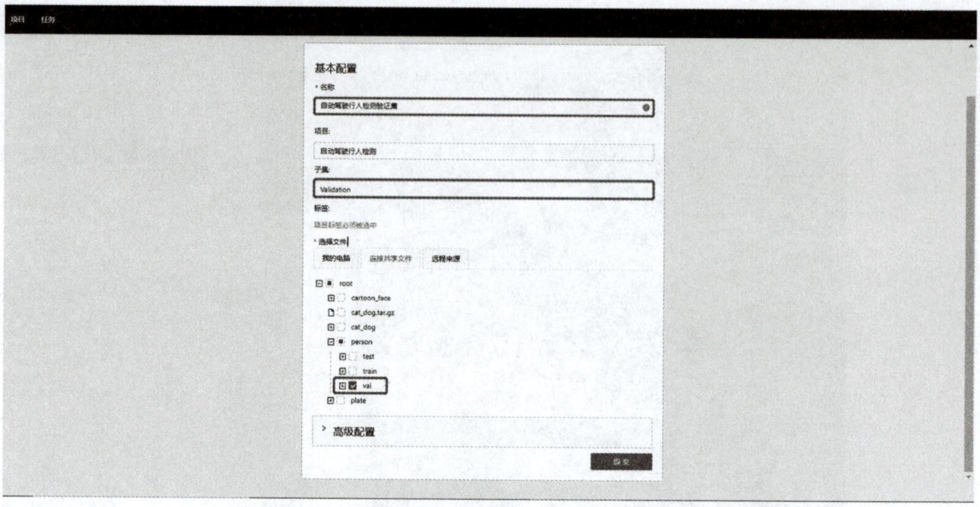

图 4.10 创建验证任务

图 4.11 进入验证集标注作业

(2)将一张图片中的对象标注完成后,单击上方工具栏中的"下一帧"按钮。
(3)继续标注,直至整个数据集标注完成。

步骤 7　导出标注验证集

(1)选择"菜单"→"导出为数据集"→"TFRecord 1.0"命令。
(2)标注完成的数据集导出后是一个压缩文件,保存在浏览器默认的下载路径中。
(3)将该文件解压缩后会得到 default.tfrecord 和 label_map.pbtxt 文件,把 default.tfrecord 重命名为 val.tfrecord。
(4)找到之前保存好的 train.tfrecord 文件,并将 val.tfrecord、train.tfrecord、label_map.pbtxt 三个文件存放到一起备用。

步骤 8　上传文件

（1）打开系统提供的 winSCP 工具，找到之前准备好的 val.tfrecord、train.tfrecord、label_map.pbtxt 文件，把这三个文件上传到数据处理平台中的 home/student/projects/unit4/data 目录下，如图 4.12 所示。

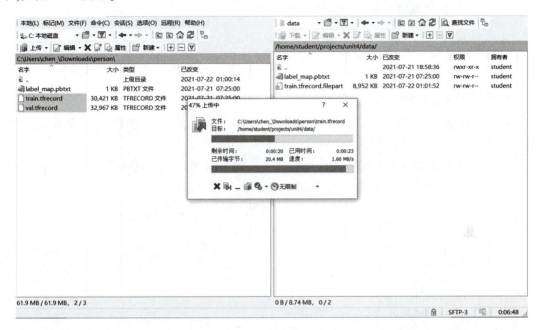

图 4.12　上传标注数据文件

（2）上传成功后，在平台上可以看到以下三个文件，如图 4.13 所示。

图 4.13　数据标注文件上传成功

任务 4　行人检测模型训练

🖥 任务描述

在本任务中，将搭建训练模型、配置预训练模型参数，对已标注的数据集进行训练，得到训练模型，并学习使用可视化的工具查看训练结果。

✏ 任务操作

步骤 1　搭建模型

在开发环境中打开准备预训练模型相关目录。

```
$ cd ~/projects/unit4
$ mkdir pretrain_models
$ cd pretrain_models
```

下载算法团队提供的预训练模型,并解压缩。

＊注意需要把地址修改成对应的资源平台地址。

```
$ wget http://172.16.33.72/dataset/dist/zy_ptm_u4.tar.gz
$ tar zxvf zy_ptm_u4.tar.gz
$ rm zy_ptm_u4.tar.gz
```

步骤2　配置训练模型

在开发环境中打开/home/student/projects/unit4/data 目录,创建训练模型配置文件person.config。

(1)主干网络配置。主干网络是整个模型训练的基础,标记了当前模型识别的物体类别等重要信息。本项目行人识别为一类,因此 num_classes 为1。

```
num_classes:1
box_coder {
  faster_rcnn_box_coder {
    y_scale:10.0
    x_scale:10.0
    height_scale:5.0
    width_scale:5.0
  }
}
matcher {
  argmax_matcher {
    matched_threshold:0.5
    unmatched_threshold:0.5
    ignore_thresholds:false
    negatives_lower_than_unmatched:true
    force_match_for_each_row:true
  }
}
similarity_calculator {
  iou_similarity {
  }
}
```

(2)先验框配置和图片分辨率配置。image_resizer 表示模型输入图片分辨率,此处为标准的 300×300 像素,因此 height 为 300,width 为 300。

```
anchor_generator {
  ssd_anchor_generator {
    num_layers:6
    min_scale:0.2
    max_scale:0.95
    aspect_ratios:1.0
```

```
      aspect_ratios:2.0
      aspect_ratios:0.5
      aspect_ratios:3.0
      aspect_ratios:0.3333
    }
  }
  image_resizer {
    fixed_shape_resizer {
      height:300
      width:300
    }
  }
```

(3)边界预测框配置。

```
box_predictor {
  convolutional_box_predictor {
    min_depth:0
    max_depth:0
    num_layers_before_predictor:0
    use_dropout:false
    dropout_keep_probability:0.8
    kernel_size:1
    box_code_size:4
    apply_sigmoid_to_scores:false
    conv_hyperparams {
      activation:RELU_6, regularizer {
        l2_regularizer {
          weight:0.00004
        }
      }
      initializer {
        truncated_normal_initializer {
          stddev:0.03
          mean:0.0
        }
      }
      batch_norm {
        train:true,
        scale:true,
        center:true,
        decay:0.9997,
        epsilon:0.001,
      }
    }
  }
}
```

(4)特征提取网络配置。

```
feature_extractor {
  type:'ssd_mobilenet_v2'
  min_depth:16
  depth_multiplier:1.0
  conv_hyperparams {
    activation:RELU_6,
    regularizer {
      l2_regularizer {
        weight:0.00004
      }
    }
    initializer {
      truncated_normal_initializer {
        stddev:0.03
        mean:0.0
      }
    }
    batch_norm {
      train:true,
      scale:true,
      center:true,
      decay:0.9997,
      epsilon:0.001,
    }
  }
}
```

(5)模型损失函数配置。

```
loss {
  classification_loss {
    weighted_sigmoid {
    }
  }
  localization_loss {
    weighted_smooth_l1 {
    }
  }
  hard_example_miner {
    num_hard_examples:3000
    iou_threshold:0.99
    loss_type:CLASSIFICATION
    max_negatives_per_positive:3
    min_negatives_per_image:3
  }
  classification_weight:1.0
  localization_weight:1.0
}
```

```
normalize_loss_by_num_matches:true
post_processing {
  batch_non_max_suppression {
    score_threshold:1e-8
    iou_threshold:0.6
    max_detections_per_class:100
    max_total_detections:100
  }
  score_converter:SIGMOID
}
```

（6）训练集数据配置。batch_size 代表批处理每次迭代的数据量，initial_learning_rate 代表初始学习率，fine_tune_checkpoint 指向预训练模型文件，input_path 指向训练集的 tfrecord 文件，label_map_path 指向标签映射文件。

```
train_config:{
  batch_size:12
  optimizer {
    rms_prop_optimizer:{
      learning_rate:{
        exponential_decay_learning_rate {
          initial_learning_rate:0.004
          decay_steps:1000
          decay_factor:0.95
        }
      }
      momentum_optimizer_value:0.9
      decay:0.9
      epsilon:1.0
    }
  }
  fine_tune_checkpoint:"pretrain_models/zy_ptm_u4/model.ckpt"
  fine_tune_checkpoint_type:"detection"
  num_steps:2000
  data_augmentation_options {
    random_horizontal_flip {
    }
  }
  data_augmentation_options {
    ssd_random_crop {
    }
  }
}

train_input_reader:{
  tf_record_input_reader {
    input_path:"data/train.tfrecord"
  }
  label_map_path:"data/label_map.pbtxt"
}
```

（7）验证集数据配置。num_examples 代表验证集样本数量，input_path 指向验证集的 tfrecord 文件，label_map_path 指向标签映射文件。

```
eval_config:{
  num_examples:50
  max_evals:1
}
eval_input_reader:{
  tf_record_input_reader {
    input_path:"data/val.tfrecord"
  }
  label_map_path:"data/label_map.pbtxt"
  shuffle:false
  num_readers:1
}
```

模型配置文件 person.config 的完整源代码文件参见 U4-person.config.pdf。

完整源代码

person.config

步骤3　创建训练文件

在开发环境中打开/home/student/projects/unit4/目录，创建训练程序 train.py。

（1）导入训练所需模块和函数。

```
import functools
import json
import os
import tensorflow as tf
from object_detection.builders import dataset_builder
from object_detection.builders import graph_rewriter_builder
from object_detection.builders import model_builder
from object_detection.legacy import trainer
from object_detection.utils import config_util
```

（2）定义输入参数。

```
os.environ["TF_CPP_MIN_LOG_LEVEL"] = '3'
tf.logging.set_verbosity(tf.logging.INFO)
flags = tf.app.flags
flags.DEFINE_string('master','','')
flags.DEFINE_integer('task',0,'task id')
flags.DEFINE_integer('num_clones',1,'')
```

```python
flags.DEFINE_boolean('clone_on_cpu',False,'')
flags.DEFINE_integer('worker_replicas',1,'')
flags.DEFINE_integer('ps_tasks',0,'')
flags.DEFINE_string('train_dir','','Directory to save the checkpoints and training summaries.')
flags.DEFINE_string('pipeline_config_path','','Path to a pipeline config.')
flags.DEFINE_string('train_config_path','','Path to a train_pb2.TrainConfig.')
flags.DEFINE_string('input_config_path','','Path to an input_reader_pb2.InputReader.')
flags.DEFINE_string('model_config_path','','Path to a model_pb2.DetectionModel.')
FLAGS = flags.FLAGS
```

(3)训练主函数:加载模型配置。

```python
@tf.contrib.framework.deprecated(None,'Use object_detection/model_main.py.')
def main(_):
  assert FLAGS.train_dir,''train_dir' is missing.'
  if FLAGS.task = = 0:tf.gfile.MakeDirs(FLAGS.train_dir)
  if FLAGS.pipeline_config_path:
    configs = config_util.get_configs_from_pipeline_file(
        FLAGS.pipeline_config_path)
    if FLAGS.task = = 0:
      tf.gfile.Copy(FLAGS.pipeline_config_path,
                    os.path.join(FLAGS.train_dir,'pipeline.config'),
                    overwrite = True)
  else:
    configs = config_util.get_configs_from_multiple_files(
        model_config_path = FLAGS.model_config_path,
        train_config_path = FLAGS.train_config_path,
        train_input_config_path = FLAGS.input_config_path)
    if FLAGS.task = = 0:
      for name,config in [('model.config',FLAGS.model_config_path),
                          ('train.config',FLAGS.train_config_path),
                          ('input.config',FLAGS.input_config_path)]:
        tf.gfile.Copy(config,os.path.join(FLAGS.train_dir,name),
                      overwrite = True)

  model_config = configs['model']
  train_config = configs['train_config']
  input_config = configs['train_input_config']
  model_fn = functools.partial(
      model_builder.build,
      model_config = model_config,
      is_training = True)
```

(4)训练主函数:设计模型线程和迭代循环。

```python
def get_next(config):
  return dataset_builder.make_initializable_iterator(
      dataset_builder.build(config)).get_next()
```

```python
create_input_dict_fn = functools.partial(get_next,input_config)

env = json.loads(os.environ.get('TF_CONFIG','{}'))
cluster_data = env.get('cluster',None)
cluster = tf.train.ClusterSpec(cluster_data) if cluster_data else None
task_data = env.get('task',None) or {'type':'master','index':0}
task_info = type('TaskSpec',(object,),task_data)

ps_tasks = 0
worker_replicas = 1
worker_job_name = 'lonely_worker'
task = 0
is_chief = True
master = ''

if cluster_data and 'worker' in cluster_data:
    worker_replicas = len(cluster_data['worker']) + 1
if cluster_data and 'ps' in cluster_data:
    ps_tasks = len(cluster_data['ps'])

if worker_replicas > 1 and ps_tasks < 1:
    raise ValueError('At least 1 ps task is needed for distributed training.')

if worker_replicas >= 1 and ps_tasks > 0:
    server = tf.train.Server(tf.train.ClusterSpec(cluster),protocol = 'grpc',
                             job_name = task_info.type,
                             task_index = task_info.index)
    if task_info.type == 'ps':
        server.join()
        return
    worker_job_name = '%s/task:%d' % (task_info.type,task_info.index)
    task = task_info.index
    is_chief = (task_info.type == 'master')
    master = server.target
```

(5)训练主函数:记录训练日志,配置训练函数参数。

```python
graph_rewriter_fn = None
if 'graph_rewriter_config' in configs:
    graph_rewriter_fn = graph_rewriter_builder.build(
        configs['graph_rewriter_config'],is_training = True)

trainer.train(
    create_input_dict_fn,
    model_fn,
    train_config,
    master,
    task,
    FLAGS.num_clones,
```

```
            worker_replicas,
            FLAGS.clone_on_cpu,
            ps_tasks,
            worker_job_name,
            is_chief,
            FLAGS.train_dir,
            graph_hook_fn=graph_rewriter_fn)
    print("模型训练完成!")

if __name__=='__main__':
    tf.app.run()
```

训练程序 train.py 的完整源代码文件参见 U4-train.py.pdf。

train.py

步骤4　训练模型

运行训练程序 train.py，如图 4.14 所示，程序读取配置文件 Person.config 中定义的训练模型、训练参数、数据集，把训练日志和检查点保存到 checkpoint 目录中。

```
$ conda activate unit4
$ python train.py --logtostderr --train_dir checkpoint --pipeline_config_path data/person.config
```

图 4.14　训练模型

步骤5　可视化训练过程

在训练过程中打开 tensorboard 可以查看训练日志,如图 4.15 所示。训练日志中记录了模型分类损失、回归损失和总损失量的变化,通过 Losses 选项中的图表可以看到训练过程中的损失在不断变化,越到后面损失越小,说明模型对训练数据的拟合度越来越高。

＊注意需要把地址修改成对应的数据处理服务器地址,然后在浏览器中输入对应地址和端口号查看。

```
$ tensorboard --host 172.16.33.11 --port 8889 --logdir checkpoint/
```

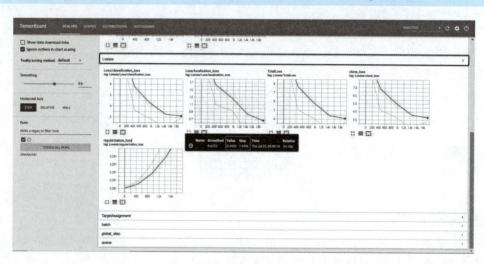

图 4.15　查看训练日志

步骤6　查看训练结果

进入 checkpoint 子目录,可以看到生成了多组模型文件,如图 4.16 所示。其中:
- model.ckpt-xxxx.meta 文件保存了计算图也就是神经网络的结构;
- model.ckpt-xxxx.data-xxxx 文件保存了模型的变量;
- model.ckpt-xxxx.index 文件保存了神经网络索引映射文件。

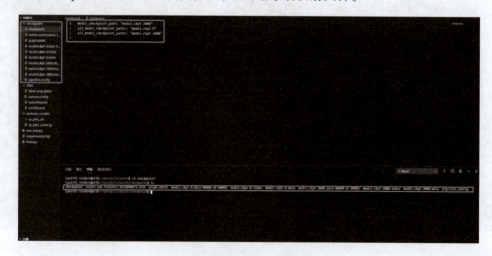

图 4.16　生成多组模型文件

项目 4　自动驾驶行人检测

任务 5　行人检测模型评估

任务描述

在本任务中,将对训练模型进行评估,判断模型的可用性。

任务操作

步骤 1　创建评估文件

在开发环境中打开/home/student/projects/unit4/目录,创建评估程序 eval.py。

(1)导入模型的各个模块并定义输入参数。

```
import functools
import os
import tensorflow as tf

from object_detection.builders import dataset_builder
from object_detection.builders import graph_rewriter_builder
from object_detection.builders import model_builder
from object_detection.legacy import evaluator
from object_detection.utils import config_util
from object_detection.utils import label_map_util

os.environ["TF_CPP_MIN_LOG_LEVEL"] = '3'
tf.compat.v1.logging.set_verbosity(tf.compat.v1.logging.ERROR)
flags = tf.app.flags
flags.DEFINE_boolean('eval_training_data', False, '')
flags.DEFINE_string('checkpoint_dir', '', '')
flags.DEFINE_string('eval_dir', '', 'Directory to write eval summaries.')
flags.DEFINE_string('pipeline_config_path', '', 'Path to a pipeline config.')
flags.DEFINE_string('eval_config_path', '', '')
flags.DEFINE_string('input_config_path', '', '')
flags.DEFINE_string('model_config_path', '', '')
flags.DEFINE_boolean('run_once', False, '')
FLAGS = flags.FLAGS
```

(2)评估主函数:加载模型配置。

```
@tf.contrib.framework.deprecated(None, 'Use object_detection/model_main.py.')
def main(unused_argv):
    assert FLAGS.checkpoint_dir, '`checkpoint_dir` is missing.'
    assert FLAGS.eval_dir, '`eval_dir` is missing.'
    tf.gfile.MakeDirs(FLAGS.eval_dir)
    if FLAGS.pipeline_config_path:
        configs = config_util.get_configs_from_pipeline_file(
            FLAGS.pipeline_config_path)
        tf.gfile.Copy(
```

77

```
            FLAGS.pipeline_config_path,
            os.path.join(FLAGS.eval_dir,'pipeline.config'),
            overwrite=True)
    else:
        configs=config_util.get_configs_from_multiple_files(
            model_config_path=FLAGS.model_config_path,
            eval_config_path=FLAGS.eval_config_path,
            eval_input_config_path=FLAGS.input_config_path)
        for name,config in [('model.config',FLAGS.model_config_path),
                            ('eval.config',FLAGS.eval_config_path),
                            ('input.config',FLAGS.input_config_path)]:
            tf.gfile.Copy(config,os.path.join(FLAGS.eval_dir,name),overwrite=True)

    model_config=configs['model']
    eval_config=configs['eval_config']
    input_config=configs['eval_input_config']
    if FLAGS.eval_training_data:
        input_config=configs['train_input_config']

    model_fn=functools.partial(
        model_builder.build,model_config=model_config,is_training=False)
```

(3) 评估主函数:定义评估循环,并记录评估日志。

```
    def get_next(config):
        return dataset_builder.make_initializable_iterator(
            dataset_builder.build(config)).get_next()

    create_input_dict_fn=functools.partial(get_next,input_config)

    categories=label_map_util.create_categories_from_labelmap(
        input_config.label_map_path)

    if FLAGS.run_once:
        eval_config.max_evals=1

    graph_rewriter_fn=None
    if 'graph_rewriter_config' in configs:
        graph_rewriter_fn=graph_rewriter_builder.build(
            configs['graph_rewriter_config'],is_training=False)
```

(4) 评估主函数:配置评估函数参数。

```
    evaluator.evaluate(
        create_input_dict_fn,
        model_fn,
        eval_config,
        categories,
        FLAGS.checkpoint_dir,
        FLAGS.eval_dir,
```

```
    graph_hook_fn = graph_rewriter_fn)
  print("模型评估完成!")

if __name__ == '__main__':
  tf.app.run()
```

评估程序 eval.py 的完整源代码文件参见 U4-eval.py.pdf。

完整源代码

eval.py

步骤 2　评估已训练模型

运行评估程序 eval.py，如图 4.17 所示，程序读取配置文件 Person.config 中定义的训练模型、训练参数、数据集，读取 checkpoint 目录中的训练结果，把评估结果保存到 evaluation 目录中。在评估过程中，可以看到对不同类别的评估结果。

```
$ conda activate unit4
$ python eval.py --logtostderr --checkpoint_dir checkpoint --eval_dir evaluation --pipeline_config_path data/person.config
```

图 4.17　评估模型

步骤 3　查看评估结果

利用 tensorboard 工具查看评估结果，如图 4.18 所示。

＊注意需要把地址修改成对应的数据处理服务器地址,然后在浏览器中输入对应地址和端口号查看。

```
$ tensorboard --host 172.16.33.11 --port 8889 --logdir evaluation/
```

步骤4　分析模型可用性

在浏览器中查看各类别的平均精确度(AP)值,越接近1说明模型的可用性越高。此时图上显示,step是2k,说明该模型是训练到2 000步时保存下来的,对应model.ckpt-2000训练模型。

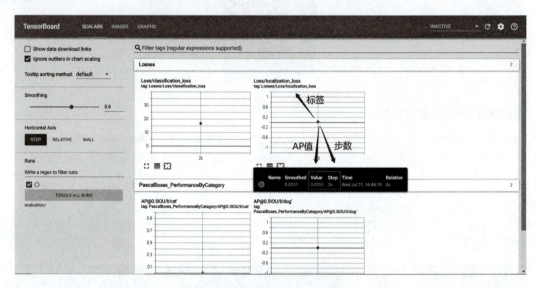

图4.18　查看评估结果

任务6　行人检测模型测试

任务描述

在本任务中,将把已经评估为可用性较强的模型,导出成为可测试的冻结图模型,用测试数据进行测试。

任务操作

步骤1　创建导出文件

在开发环境中打开/home/student/projects/unit4/目录,创建导出程序export_fz.py。

(1)导入模型转换模块,定义输入参数。

```
import os
import tensorflow as tf
from google.protobuf import text_format
from object_detection import exporter
```

```python
from object_detection.protos import pipeline_pb2

os.environ["TF_CPP_MIN_LOG_LEVEL"] = '3'
tf.compat.v1.logging.set_verbosity(tf.compat.v1.logging.ERROR)
slim = tf.contrib.slim
flags = tf.app.flags

flags.DEFINE_string('input_type','image_tensor','')
flags.DEFINE_string('input_shape',None,'[None,None,None,3]')
flags.DEFINE_string('pipeline_config_path',None,'Path to a pipeline config.')
flags.DEFINE_string('trained_checkpoint_prefix',None,'path/to/model.ckpt')
flags.DEFINE_string('output_directory',None,'Path to write outputs.')
flags.DEFINE_string('config_override','','')
flags.DEFINE_boolean('write_inference_graph',False,'')
tf.app.flags.mark_flag_as_required('pipeline_config_path')
tf.app.flags.mark_flag_as_required('trained_checkpoint_prefix')
tf.app.flags.mark_flag_as_required('output_directory')
FLAGS = flags.FLAGS
```

（2）转换模型主函数，调用模型转换函数。

```python
def main(_):
    pipeline_config = pipeline_pb2.TrainEvalPipelineConfig()
    with tf.gfile.GFile(FLAGS.pipeline_config_path,'r') as f:
        text_format.Merge(f.read(),pipeline_config)
    text_format.Merge(FLAGS.config_override,pipeline_config)
    if FLAGS.input_shape:
        input_shape = [
            int(dim) if dim != '-1' else None
            for dim in FLAGS.input_shape.split(',')
        ]
    else:
        input_shape = None
    exporter.export_inference_graph(
        FLAGS.input_type,pipeline_config,FLAGS.trained_checkpoint_prefix,
        FLAGS.output_directory,input_shape=input_shape,
        write_inference_graph=FLAGS.write_inference_graph)
    print("模型转换完成!")

if __name__ == '__main__':
    tf.app.run()
```

导出程序 export_fz.py 的完整源代码文件参见 U4-export_fz.py.pdf。

完整源代码

export_fz.py

步骤2 导出冻结图模型

运行导出程序 export_fz.py，如图 4.19 所示，程序将读取配置文件 Person.config 中定义的配置，读取 checkpoint 目录中的 model.ckpt-2000 训练模型，导出为冻结图模型，并保存到 frozen_models 目录中。

```
$ conda activate unit4
$ python export_fz.py --input_type image_tensor --pipeline_config_path data/person.config --trained_checkpoint_prefix checkpoint/model.ckpt-2000 --output_directory frozen_models
```

图 4.19 导出冻结图模型

步骤3 创建测试文件

在开发环境中打开 /home/student/projects/unit4/ 目录，创建测试文件 detect.py
(1) 导入测试所需模块和可视化函数，定义输入参数。

```
import numpy as np
import os
import tensorflow as tf
import matplotlib.pyplot as plt
from PIL import Image
from object_detection.utils import label_map_util
from object_detection.utils import visualization_utils as vis_util
from object_detection.utils import ops as utils_ops

os.environ["TF_CPP_MIN_LOG_LEVEL"] = '3'
tf.compat.v1.logging.set_verbosity(tf.compat.v1.logging.ERROR)
detect_img = '/home/student/data/person/test/5.jpg'
result_img = '/home/student/projects/unit4/img/5_result.jpg'
MODEL_NAME = 'frozen_models'
PATH_TO_FROZEN_GRAPH = MODEL_NAME + '/frozen_inference_graph.pb'
PATH_TO_LABELS = 'data/label_map.pbtxt'
```

（2）加载模型计算图和数据标签。

```
detection_graph = tf.Graph()
with detection_graph.as_default():
    od_graph_def = tf.compat.v1.GraphDef()
    with tf.io.gfile.GFile(PATH_TO_FROZEN_GRAPH,'rb') as fid:
        serialized_graph = fid.read()
        od_graph_def.ParseFromString(serialized_graph)
        tf.import_graph_def(od_graph_def,name = '')
category_index = label_map_util.create_category_index_from_labelmap(PATH_TO_LABELS,
        use_display_name = True)
```

（3）图片数据转换函数。

```
def load_image_into_numpy_array(image):
    (im_width,im_height) = image.size
    return np.array(image.getdata()).reshape((im_height,im_width,3)).astype(np.uint8)
```

（4）单张图像检测函数。

```
def run_inference_for_single_image(image,graph):
    with graph.as_default():
        with tf.compat.v1.Session() as sess:
            ops = tf.compat.v1.get_default_graph().get_operations()
            all_tensor_names = {output.name for op in ops for output in op.outputs}
            tensor_dict = {}
            for key in ['num_detections','detection_boxes','detection_scores',
                'detection_classes','detection_masks']:
                tensor_name = key + ':0'
                if tensor_name in all_tensor_names:
                    tensor_dict[key] = tf.compat.v1.get_default_graph().get_tensor_by_name(tensor_name)
            if 'detection_masks' in tensor_dict:
                detection_boxes = tf.squeeze(tensor_dict['detection_boxes'],[0])
                detection_masks = tf.squeeze(tensor_dict['detection_masks'],[0])
                real_num_detection = tf.cast(tensor_dict['num_detections'][0],tf.int32)
                detection_boxes = tf.slice(detection_boxes,[0,0],[real_num_detection,-1])
                detection_masks = tf.slice(detection_masks,[0,0,0],[real_num_detection,-1,-1])
                detection_masks_reframed = utils_ops.reframe_box_masks_to_image_masks(
                    detection_masks,detection_boxes,image.shape[1],image.shape[2])
                detection_masks_reframed = tf.cast(tf.greater
                    (detection_masks_reframed,0.5),tf.uint8)
                tensor_dict['detection_masks'] = tf.expand_dims(detection_masks_reframed,0)
            image_tensor = tf.compat.v1.get_default_graph().get_tensor_by_name('image_tensor:0')

            output_dict = sess.run(tensor_dict,feed_dict = {image_tensor:image})

            output_dict['num_detections'] = int(output_dict['num_detections'][0])
            output_dict['detection_classes'] = output_dict['detection_classes'][0].astype(np.int64)
```

```python
        output_dict['detection_boxes'] = output_dict['detection_boxes'][0]
        output_dict['detection_scores'] = output_dict['detection_scores'][0]
        if 'detection_masks' in output_dict:
            output_dict['detection_masks'] = output_dict['detection_masks'][0]
    return output_dict
```

(5) 输入图片数据,检测输入数据,保存检测结果图。

```python
image = Image.open(detect_img)
image_np = load_image_into_numpy_array(image)
#[1,None,None,3]
image_np_expanded = np.expand_dims(image_np, axis=0)
output_dict = run_inference_for_single_image(image_np_expanded, detection_graph)
vis_util.visualize_boxes_and_labels_on_image_array(
    image_np,
    output_dict['detection_boxes'],
    output_dict['detection_classes'],
    output_dict['detection_scores'],
    category_index,
    instance_masks = output_dict.get('detection_masks'),
    use_normalized_coordinates = True,
    line_thickness = 6)
plt.figure()
plt.axis('off')
plt.imshow(image_np)
plt.savefig(result_img, bbox_inches = 'tight', pad_inches = 0)
print("测试%s完成,结果保存在%s" %(detect_img, result_img))
```

测试程序 detect.py 的完整源代码文件参见 U4-detect.py.pdf。

完整源代码

detect.py

步骤4 测试并查看结果

创建 img 目录存放测试结果,运行测试程序 detect.py,如图4.20所示,并查看输出的结果图片,如图4.21所示。

```
$ mkdir img
$ python detect.py
```

图 4.20　测试模型

图 4.21 查看测试结果

任务 7　行人检测模型部署

在本任务中,将把经过测试确认可用的模型,转换成为应用部署支持的格式,然后将模型文件部署到边缘计算设备上,并实现人工智能应用的集成。

步骤 1　创建导出程序

在开发环境中打开 /home/student/projects/unit4/ 目录,创建导出程序 export_pb.py。
(1)导入模型转换模块,定义输入参数。

```
import os
import tensorflow as tf
from google.protobuf import text_format
from object_detection import export_tflite_ssd_graph_lib
from object_detection.protos import pipeline_pb2

os.environ["TF_CPP_MIN_LOG_LEVEL"] = '3'
tf.compat.v1.logging.set_verbosity(tf.compat.v1.logging.ERROR)
flags = tf.app.flags
flags.DEFINE_string('output_directory',None,'Path to write outputs.')
flags.DEFINE_string('pipeline_config_path',None,'')
flags.DEFINE_string('trained_checkpoint_prefix',None,'Checkpoint prefix.')
flags.DEFINE_integer('max_detections',10,'')
flags.DEFINE_integer('max_classes_per_detection',1,'')
flags.DEFINE_integer('detections_per_class',100,'')
```

```
flags.DEFINE_bool('add_postprocessing_op',True,'')
flags.DEFINE_bool('use_regular_nms',False,'')
flags.DEFINE_string('config_override','','')
FLAGS = flags.FLAGS
```

(2) 调用模型转换函数,完成模型转换。

```
def main(argv):
  flags.mark_flag_as_required('output_directory')
  flags.mark_flag_as_required('pipeline_config_path')
  flags.mark_flag_as_required('trained_checkpoint_prefix')

  pipeline_config = pipeline_pb2.TrainEvalPipelineConfig()

  with tf.gfile.GFile(FLAGS.pipeline_config_path,'r') as f:
    text_format.Merge(f.read(),pipeline_config)
  text_format.Merge(FLAGS.config_override,pipeline_config)
  export_tflite_ssd_graph_lib.export_tflite_graph(
      pipeline_config,FLAGS.trained_checkpoint_prefix,FLAGS.output_directory,
      FLAGS.add_postprocessing_op,FLAGS.max_detections,
      FLAGS.max_classes_per_detection,FLAGS.use_regular_nms)
  print("模型转换完成!")

if __name__ == '__main__':
  tf.app.run(main)
```

导出程序 export_pb.py 的完整源代码文件参见 U4-export_pb.py.pdf。

完整源代码

export_pb.py

步骤2 导出pb文件

运行导出程序 export_pb.py,读取配置文件 person.config 中定义的参数,读取 checkpoint 目录中的训练结果,把 tflite_pb 模型图保存到 tflite_models 目录中。

```
$ conda activate unit4
$ python export_pb.py --pipeline_config_path data/person.config --trained_checkpoint_prefix checkpoint/model.ckpt-2000 --output_directory tflite_models
```

步骤3 创建转换程序

在开发环境中打开/home/student/projects/unit4/目录,创建文件 pb_to_tflite.py。

(1) 导入模块,定义输入参数。

```
import os
import tensorflow as tf
```

```python
os.environ["TF_CPP_MIN_LOG_LEVEL"] = '3'
tf.compat.v1.logging.set_verbosity(tf.compat.v1.logging.ERROR)
flags = tf.app.flags
flags.DEFINE_string('pb_path','tflite_models/tflite_graph.pb','tflite pb file.')
flags.DEFINE_string('tflite_path','tflite_models/zy_ssd.tflite','output tflite.')
FLAGS = flags.FLAGS
```

(2) 转换为 tflite 模型。

```python
def convert_pb_to_tflite(pb_path,tflite_path):
    # 模型输入节点
    input_tensor_name = ["normalized_input_image_tensor"]
    input_tensor_shape = {"normalized_input_image_tensor":[1,300,300,3]}
    # 模型输出节点
    classes_tensor_name = ['TFLite_Detection_PostProcess','TFLite_Detection_PostProcess:1','TFLite_Detection_PostProcess:2','TFLite_Detection_PostProcess:3']
    # 转换为 tflite 模型
    converter = tf.lite.TFLiteConverter.from_frozen_graph(pb_path,
                                                          input_tensor_name,
                                                          classes_tensor_name,
                                                          input_tensor_shape)
    converter.allow_custom_ops = True
    converter.optimizations = [tf.lite.Optimize.DEFAULT]
    tflite_model = converter.convert()
```

(3) tflite 模型写入。

```python
    converter.allow_custom_ops = True
    converter.optimizations = [tf.lite.Optimize.DEFAULT]
    tflite_model = converter.convert()
    # 模型写入
    if not tf.gfile.Exists(os.path.dirname(tflite_path)):
        tf.gfile.MakeDirs(os.path.dirname(tflite_path))
    with open(tflite_path,"wb") as f:
        f.write(tflite_model)
    print("Save tflite model at %s" % tflite_path)
    print("模型转换完成!")

if __name__ == '__main__':
    convert_pb_to_tflite(FLAGS.pb_path,FLAGS.tflite_path)
```

转换程序 pb_to_tflite.py 的完整源代码文件参见 U4-pb_to_tflite.py.pdf。

完整源代码

pb_to_tflite.py

步骤4 转换 tflite 文件

运行文件程序 pb_to_tflite.py。

```
$ python pb_to_tflite.py
```

步骤5 创建推理执行程序

在开发环境中打开 /home/student/projects/unit4/tflite_models 目录，创建推理执行程序 func_detection_img.py。

(1) 导入所用模块。

```python
import os
import cv2
import numpy as np
import sys
import glob
import importlib.util
import base64
```

(2) 定义模型和数据推理器。

```python
def update_image(image_data,GRAPH_NAME='zy_ssd.tflite',min_conf_threshold=0.5,
            use_TPU=False,model_dir='util'):
    from tflite_runtime.interpreter import Interpreter
    CWD_PATH=os.getcwd()
    PATH_TO_CKPT=os.path.join(CWD_PATH,model_dir,GRAPH_NAME)

    labels=['person']

    interpreter=Interpreter(model_path=PATH_TO_CKPT)

    interpreter.allocate_tensors()

    input_details=interpreter.get_input_details()
    output_details=interpreter.get_output_details()
    height=input_details[0]['shape'][1]
    width=input_details[0]['shape'][2]

    floating_model=(input_details[0]['dtype']==np.float32)

    input_mean=127.5
    input_std=127.5
```

(3) 输入图像并转换图像数据为张量。

```python
# base64 解码
img_data=base64.b64decode(image_data)
# 转换为 np 数组
img_array=np.fromstring(img_data,np.uint8)
# 转换成 opencv 可用格式
image=cv2.imdecode(img_array,cv2.COLOR_RGB2BGR)
```

```
image_rgb = cv2.cvtColor(image,cv2.COLOR_BGR2RGB)
imH,imW,_ = image.shape
image_resized = cv2.resize(image_rgb,(width,height))
input_data = np.expand_dims(image_resized,axis = 0)

if floating_model:
    input_data = (np.float32(input_data)-input_mean)/input_std

interpreter.set_tensor(input_details[0]['index'],input_data)
interpreter.invoke()

boxes = interpreter.get_tensor(output_details[0]['index'])[0]
classes = interpreter.get_tensor(output_details[1]['index'])[0]
scores = interpreter.get_tensor(output_details[2]['index'])[0]
```

(4)检测图片,并可视化输出结果。

```
for i in range(len(scores)):
    if((scores[i] >min_conf_threshold) and (scores[i] < =1.0)):
        ymin = int(max(1,(boxes[i][0] * imH)))
        xmin = int(max(1,(boxes[i][1] * imW)))
        ymax = int(min(imH,(boxes[i][2] * imH)))
        xmax = int(min(imW,(boxes[i][3] * imW)))

        cv2.rectangle(image,(xmin,ymin),(xmax,ymax),(10,255,0),2)

        object_name = labels[int(classes[i])]
        label = '% s:% d% % ' % (object_name,int(scores[i] * 100))
        labelSize,baseLine = cv2.getTextSize(label,cv2.FONT_HERSHEY_SIMPLEX,0.7,2)
        label_ymin = max(ymin,labelSize[1] +10)
        cv2.rectangle(image,(xmin,label_ymin-labelSize[1]-10),
            (xmin + labelSize[0],label_ymin +baseLine-10),(255,255,255),cv2.FILLED)
        cv2.putText(image,label,(xmin,label_ymin-7),cv2.FONT_HERSHEY_SIMPLEX,0.7,
(0,0,0),2)

    image_bytes = cv2.imencode('.jpg',image)[1].tostring()
    image_base64 = base64.b64encode(image_bytes).decode()
    return image_base64
```

推理执行程序 func_detection_img.py 的完整源代码文件参见 U4-func_detection_img.py.pdf。

完整源代码

func_detection_img.py

步骤6　部署到边缘设备

（1）把模型 zy_ssd.tflite 文件、推理执行程序 func_detection_img.py 文件复制到边缘计算设备中。

＊注意把 IP 地址修改成对应的推理机地址。

```
$ scp tflite_models/zy_ssd.tflite student@172.16.33.118:/home/student/zy-panel-check/util/
$ scp tflite_models/func_detection_img.py student@172.16.33.118:/home/student/zy-panel-check/util/
```

（2）通过平台上的"应用部署"按钮，上传或输入图片 URL 检测，应用部署成功结果如图 4.22 所示。

图 4.22　应用部署成功

项目小结

通过亲自动手完成数据标注、数据训练、模型导出等任务，你实现了一个以人为目标的机器识别模型，并部署到边缘计算设备上。同时，通过学习，你也了解到做自动驾驶是一个很艰难的旅程，不断地解决问题之后又会出现新的问题，不过正是因为过程的艰难，才带来更多的快乐。祝贺你，已经做好步入新的工作岗位的心理准备。

多学一点：视觉传感器是汽车自动驾驶领域常用的一种传感器，与其他传感器相比，具有检测信息量大、性价比高等优点。一般的检测步骤包括图像获取、预处理、目标分类、目标定位跟踪等。为了实现真正意义上的行人检测，还需要在本身运动的情况下，获知前方或者侧前方的行人信息包括行走趋势分析的结果信息，比如行走速度、横穿加速度、纵行加速度、与本车距离等，因此必须怀着一颗敬畏的心，从事这项有意义的工作。祝愿你在未来的学习中掌握更多的技能，在实际工作中灵活运用，成为一名优秀的工程师。

项目 5

智慧社区交通工具检测

项目背景

学电子信息的你毕业后进入了一家小有名气的系统集成公司,公司承接了多个智慧社区建设项目,也有很多样板工程。最近公司承接了一个智慧社区改造项目,该小区建造于 20 世纪 80 年代,是一个典型的老小区,基础设施相对落后,流动人口多,管理难度很大,其中停车难、乱停车问题尤为明显。小区居民表示,近年随着车辆的增加,小区车位早已不能满足居民停车需求。小区以及周边车辆乱停乱放、损毁公共设施、占用消防通道的现象日益严重。针对这个突出问题,公司拿出的解决方案是:"梳理停车需求,建立智慧停车系统,利用周边闲置空地增加车位,重新划定小区消防通道……"其中有一个需求点"针对车辆'车不入位'及'违章停车'等情况,系统会自动将相应情况即时发送到停车管理员手持终端"。针对该需求,项目经理分配给你的任务是:根据拿到的数据和预训练模型,生成一个准确率较高的图像识别模型,实现交通工具的自动识别。然后由另一组同事负责判断是否属于"车不入位"及"违章停车",并发送报警信息。

提示:交通工具的识别对于计算机视觉来说,是典型的任务之一,项目经理已经提供了数百张照片和一组预训练模型。经过分析,梳理出的需求是通过图片识别公共汽车、小汽车、自行车、摩托车 4 种交通工具,因此需要建立 4 个类别。通过数据标注、模型训练、模型导出等工作实现自动识别。

任务 1 数据准备

任务描述

在本任务中,将获得项目团队提供的交通工具图片数据集,并把这些图片数据导入人工智能数据处理平台中,分别保存在训练、验证、测试等不同的目录中,为后续数据标注工作做好准备。

步骤1 数据采集

项目经理已经准备好了相关图片数据。

步骤2 数据整理

把数据下载到工作目录,解压缩。

在终端命令行窗口中执行以下操作。

*注意第二行命令需要把地址修改成对应的资源平台地址。

```
$ cd ~/data
$ wget http://172.16.33.72/dataset/vehicle.tar.gz
$ tar zxvf vehicle.tar.gz
```

在终端命令行窗口中执行以下操作,查看解压缩后的目录,如图5.1所示。

```
$ cd ~/data/vehicle
$ ls
```

图5.1 查看解压缩后的目录

任务2　工程环境准备

在本任务中,将创建本项目的开发环境,并做基础配置,满足数据标注、模型训练、评估和部署的基础条件。

步骤1 创建工程目录

在开发环境中打开为本项目创建的工程目录,在终端命令行窗口中执行以下操作:

```
$ mkdir ~/projects/unit5
$ mkdir ~/projects/unit5/data
$ cd ~/projects/unit5
```

步骤2 创建开发环境

在Python 3.6中创建名为unit5的虚拟环境。

```
$ conda create -n unit5 python=3.6
```

输入y继续完成,然后执行以下操作激活开发环境。

```
$ conda activate unit5
```

步骤3 配置 GPU 环境

安装 tensorflow-gpu 1.15 环境。

```
$ conda install tensorflow-gpu=1.15
```

输入 y 继续完成 GPU 环境的配置,如图 5.2 所示。

图 5.2 完成 GPU 环境的配置

步骤4 配置依赖环境

(1)在开发环境中打开/home/student/projects/unit5 目录,创建依赖清单文件 requirements.txt。

(2)把以下内容写到 requirements.txt 清单文件中,然后执行命令,安装依赖库环境,完成依赖环境的配置,如图 5.3 所示。

图 5.3 完成依赖环境的配置

```
# requirements.txt
Cython
contextlib2
matplotlib
pillow
lxml
jupyter
pycocotools
click
PyYAML
joblib
autopep8

#执行以下命令
$ conda activate unit5
$ pip install -r requirements.txt
```

步骤5　配置图像识别库环境

(1) 安装中育 object_detection 库和中育 slim 库。中育 object_detection 库和中育 slim 库为目标检测模型库，后面的任务将通过这两个库生成基础的目标检测模型。

(2) 在终端命令行窗口中执行以下操作，完成后删除安装程序。

＊注意第一行和第三行命令需要把地址修改成对应的资源平台地址。

```
$ wget http://172.16.33.72/dataset/dist/zy_od_1.0.tar.gz
$ pip install zy_od_1.0.tar.gz
$ wget http://172.16.33.72/dataset/dist/zy_slim_1.0.tar.gz
$ pip install zy_slim_1.0.tar.gz
$ rm zy_od_1.0.tar.gz zy_slim_1.0.tar.gz
```

步骤6　验证环境

在终端命令行窗口中执行以下操作，查看验证结果，如图 5.4 所示。

图 5.4　验证结果

*注意需要把地址修改成对应的资源平台地址。

```
$ wget http://172.16.33.72/dataset/script/env_test.py
$ python env_test.py
```

任务 3　交通工具图片数据标注

任务描述

在本任务中,将使用图片标注工具完成数据标注,导出为数据集文件,并保存标签映射文件。

任务操作

步骤 1　添加标注标签

(1)创建名称为"智慧社区交通工具检测"标注项目,如图 5.5 所示。
(2)添加 4 类标签,分别为 bus、bicycle、motorcycle、car。
(3)注意设置为不同的颜色标签以示区分。

图 5.5　创建标注项目

步骤 2　创建训练集任务

(1)任务名称为"智慧社区交通工具检测训练集"。
(2)任务子集选择 Train。
(3)选择文件使用"连接共享文件",选中本项目任务 1 中整理的 train 子目录,如图 5.6 所示。
(4)单击"提交"按钮,完成创建。

步骤 3　标注训练集数据

(1)打开"智慧社区交通工具检测训练集",单击左下方的"作业 #"进入标注作业,如图 5.7 所示。
(2)使用加锁,可以避免对已标注对象的误操作;将一张图片中的对象标注完成后,单击上方工具栏中的"下一帧"按钮继续标注,如图 5.8 所示。

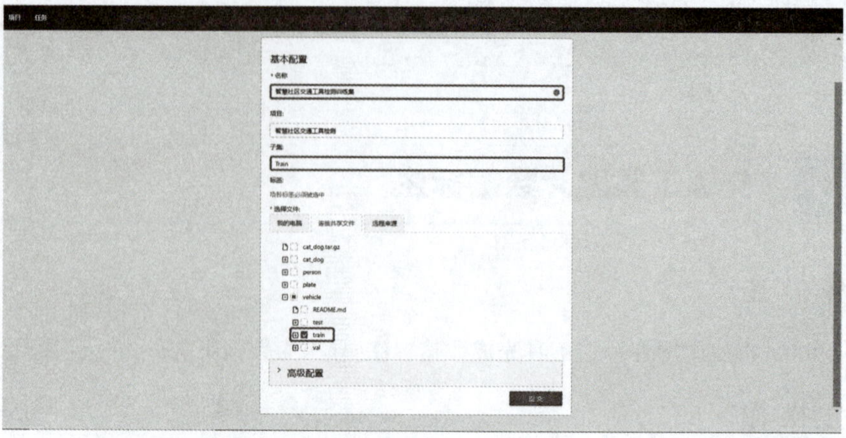

图 5.6 创建训练任务

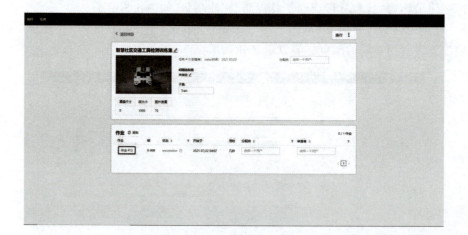

图 5.7 进入训练集标注作业

图 5.8 标注图片

(3)继续标注,直至整个数据集标注完成。

步骤4　导出标注训练集

(1)选择"菜单"→"导出为数据集"→"TFRecord 1.0"命令,如图5.9所示。

图5.9　导出为数据集

(2)标注完成的数据导出后是一个压缩包 zip 文件,保存在浏览器默认的下载路径中。
(3)将该文件解压缩,并把 default.tfrecord 重命名为 train.tfrecord。

步骤5　创建验证集任务

(1)任务名称为"智慧社区交通工具检测验证集"。
(2)任务子集选择 Validation。
(3)选择文件使用"连接共享文件",选中本项目任务1中整理的 val 子目录,如图5.10所示。
(4)单击"提交"按钮,完成创建。

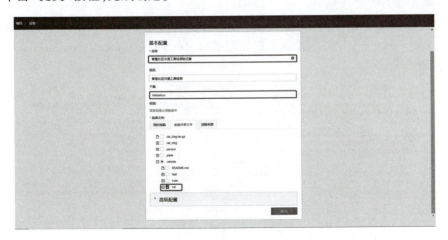

图5.10　创建验证任务

步骤6　标注验证集数据

(1)打开"智慧社区交通工具检测验证集",单击左下方的"作业#"进入标注作业,如图5.11

所示。

图 5.11　进入验证集标注作业

(2) 将一张图片中的对象标注完成后,单击上方工具栏中的"下一帧"按钮。

(3) 继续标注,直至整个数据集标注完成。

步骤 7　导出标注验证集

(1) 选择"菜单"→"导出为数据集"→"TFRecord 1.0"命令。

(2) 标注完成的数据集导出后是一个压缩文件,保存在浏览器默认的下载路径中。

(3) 将该文件解压缩后会得到 default.tfrecord 和 label_map.pbtxt 文件,把 default.tfrecord 重命名为 val.tfrecord。

(4) 找到之前保存好的 train.tfrecord 文件,并将 val.tfrecord、train.tfrecord、label_map.pbtxt 三个文件存放到一起备用。

步骤 8　上传文件

(1) 打开系统提供的 winSCP 工具,找到之前准备好的 val.tfrecord、train.tfrecord、label_map.pbtxt 文件,把这三个文件上传到数据处理平台中的 home/student/projects/unit5/data 目录下,如图 5.12 所示。

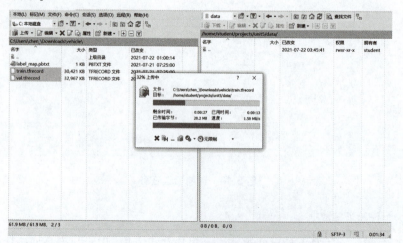

图 5.12　上传标注数据文件

(2)上传成功后,在平台上可以看到以下三个文件,如图 5.13 所示。

图 5.13　数据标注文件上传成功

任务 4　交通工具检测模型训练

在本任务中,将搭建训练模型、配置预训练模型参数,对已标注的数据集进行训练,得到训练模型,并学习使用可视化的工具查看训练结果。

步骤 1　搭建模型

(1)在开发环境中打开准备预训练模型相关目录。

```
$ cd ~/projects/unit5
$ mkdir pretrain_models
$ cd pretrain_models
```

(2)下载算法团队提供的预训练模型,并解压缩。

* 注意需要把地址修改成对应的资源平台地址。

```
$ wget http://172.16.33.72/dataset/dist/zy_ptm_u5.tar.gz
$ tar zxvf zy_ptm_u5.tar.gz
$ rm zy_ptm_u5.tar.gz
```

步骤 2　配置训练模型

在开发环境中打开/home/student/projects/unit5/data 目录,创建训练模型配置文件 vehicle.config。

(1)主干网络配置。主干网络是整个模型训练的基础,标记了当前模型识别的物体类别等重要信息。本项目交通工具识别为四类,因此 num_classes 为 4。

```
num_classes:4
box_coder {
  faster_rcnn_box_coder {
    y_scale:10.0
    x_scale:10.0
    height_scale:5.0
    width_scale:5.0
  }
}
matcher {
```

```
    argmax_matcher {
      matched_threshold:0.5
      unmatched_threshold:0.5
      ignore_thresholds:false
      negatives_lower_than_unmatched:true
      force_match_for_each_row:true
    }
  }
  similarity_calculator {
    iou_similarity {
    }
  }
```

(2)先验框配置和图片分辨率配置。image_resizer 表示模型输入图片分辨率,此处为标准的 300×300,因此 height 为 300,width 为 300。

```
  anchor_generator {
    ssd_anchor_generator {
      num_layers:6
      min_scale:0.2
      max_scale:0.95
      aspect_ratios:1.0
      aspect_ratios:2.0
      aspect_ratios:0.5
      aspect_ratios:3.0
      aspect_ratios:0.3333
    }
  }
  image_resizer {
    fixed_shape_resizer {
      height:300
      width:300
    }
  }
```

(3)边界预测框配置。

```
  box_predictor {
    convolutional_box_predictor {
      min_depth:0
      max_depth:0
      num_layers_before_predictor:0
      use_dropout:false
      dropout_keep_probability:0.8
      kernel_size:1
      box_code_size:4
      apply_sigmoid_to_scores:false
      conv_hyperparams {
        activation:RELU_6,
        regularizer {
```

```
      l2_regularizer {
        weight:0.00004
      }
    }
    initializer {
      truncated_normal_initializer {
        stddev:0.03
        mean:0.0
      }
    }
    batch_norm {
      train:true,
      scale:true,
      center:true,
      decay:0.9997,
      epsilon:0.001,
    }
  }
 }
}
```

(4)特征提取网络配置。

```
feature_extractor {
  type:'ssd_mobilenet_v2'
  min_depth:16
  depth_multiplier:1.0
  conv_hyperparams {
    activation:RELU_6,
    regularizer {
      l2_regularizer {
        weight:0.00004
      }
    }
    initializer {
      truncated_normal_initializer {
        stddev:0.03
        mean:0.0
      }
    }
    batch_norm {
      train:true,
      scale:true,
      center:true,
      decay:0.9997,
      epsilon:0.001,
    }
  }
}
```

（5）模型损失函数配置。

```
loss {
  classification_loss {
    weighted_sigmoid {
    }
  }
  localization_loss {
    weighted_smooth_l1 {
    }
  }
  hard_example_miner {
    num_hard_examples:3000
    iou_threshold:0.99
    loss_type:CLASSIFICATION
    max_negatives_per_positive:3
    min_negatives_per_image:3
  }
  classification_weight:1.0
  localization_weight:1.0
}
normalize_loss_by_num_matches:true
post_processing {
  batch_non_max_suppression {
    score_threshold:1e-8
    iou_threshold:0.6
    max_detections_per_class:100
    max_total_detections:100
  }
  score_converter:SIGMOID
}
```

（6）训练集数据配置。batch_size 代表批处理每次迭代的数据量，initial_learning_rate 代表初始学习率，fine_tune_checkpoint 指向预训练模型文件，input_path 指向训练集的 tfrecord 文件，label_map_path 指向标签映射文件。

```
train_config:{
  batch_size:12
  optimizer {
    rms_prop_optimizer:{
      learning_rate:{
        exponential_decay_learning_rate {
          initial_learning_rate:0.004
          decay_steps:1000
          decay_factor:0.95
        }
      }
      momentum_optimizer_value:0.9
      decay:0.9
```

```
    epsilon:1.0
  }
}
fine_tune_checkpoint:"pretrain_models/zy_ptm_u5/model.ckpt"
fine_tune_checkpoint_type:  "detection"
num_steps:2000
data_augmentation_options {
  random_horizontal_flip {
  }
}
data_augmentation_options {
  ssd_random_crop {
  }
}

train_input_reader:{
  tf_record_input_reader {
    input_path:"data/train.tfrecord"
  }
  label_map_path:"data/label_map.pbtxt"
}
```

（7）验证集数据配置。num_examples 代表验证集样本数量，input_path 指向验证集的 tfrecord 文件，label_map_path 指向标签映射文件。

```
eval_config:{
  num_examples:50
  max_evals:1
}

eval_input_reader:{
  tf_record_input_reader {
    input_path:"data/val.tfrecord"
  }
  label_map_path:"data/label_map.pbtxt"
  shuffle:false
  num_readers:1
}
```

模型配置文件 vehicle.config 的完整源代码文件参见 U5-vehicle.config.pdf。

完整源代码

vehicle.config

步骤3 创建训练文件

在开发环境中打开/home/student/projects/unit5/目录,创建训练程序 train.py。

(1)导入训练所需模块和函数。

```python
import functools
import json
import os
import tensorflow as tf
from object_detection.builders import dataset_builder
from object_detection.builders import graph_rewriter_builder
from object_detection.builders import model_builder
from object_detection.legacy import trainer
from object_detection.utils import config_util
```

(2)定义输入参数。

```python
os.environ["TF_CPP_MIN_LOG_LEVEL"] = '3'
tf.logging.set_verbosity(tf.logging.INFO)
flags = tf.app.flags
flags.DEFINE_string('master','','')
flags.DEFINE_integer('task',0,'task id')
flags.DEFINE_integer('num_clones',1,'')
flags.DEFINE_boolean('clone_on_cpu',False,'')
flags.DEFINE_integer('worker_replicas',1,'')
flags.DEFINE_integer('ps_tasks',0,'')
flags.DEFINE_string('train_dir','','Directory to save the checkpoints and training summaries.')
flags.DEFINE_string('pipeline_config_path','','Path to a pipeline config.')
flags.DEFINE_string('train_config_path','','Path to a train_pb2.TrainConfig.')
flags.DEFINE_string('input_config_path','','Path to an input_reader_pb2.InputReader.')
flags.DEFINE_string('model_config_path','','Path to a model_pb2.DetectionModel.')
FLAGS = flags.FLAGS
```

(3)训练主函数:加载模型配置。

```python
@tf.contrib.framework.deprecated(None,'Use object_detection/model_main.py.')
def main(_):
    assert FLAGS.train_dir,''train_dir' is missing.'
    if FLAGS.task==0:tf.gfile.MakeDirs(FLAGS.train_dir)
    if FLAGS.pipeline_config_path:
        configs=config_util.get_configs_from_pipeline_file(
            FLAGS.pipeline_config_path)
        if FLAGS.task==0:
            tf.gfile.Copy(FLAGS.pipeline_config_path,
                os.path.join(FLAGS.train_dir,'pipeline.config'),
                overwrite=True)
    else:
        configs=config_util.get_configs_from_multiple_files(
            model_config_path=FLAGS.model_config_path,
            train_config_path=FLAGS.train_config_path,
```

```
            train_input_config_path = FLAGS.input_config_path)
    if FLAGS.task = = 0:
      for name,config in [('model.config',FLAGS.model_config_path),
                          ('train.config',FLAGS.train_config_path),
                          ('input.config',FLAGS.input_config_path)]:
        tf.gfile.Copy(config,os.path.join(FLAGS.train_dir,name), overwrite = True)

  model_config = configs['model']
  train_config = configs['train_config']
  input_config = configs['train_input_config']
  model_fn = functools.partial(
      model_builder.build,
      model_config = model_config,
      is_training = True)
```

(4)训练主函数：设计模型线程和迭代循环。

```
  def get_next(config):
      return dataset_builder.make_initializable_iterator(
          dataset_builder.build(config)).get_next()

  create_input_dict_fn = functools.partial(get_next,input_config)

  env = json.loads(os.environ.get('TF_CONFIG','{}'))
  cluster_data = env.get('cluster',None)
  cluster = tf.train.ClusterSpec(cluster_data) if cluster_data else None
  task_data = env.get('task',None) or {'type':'master','index':0}
  task_info = type('TaskSpec',(object,),task_data)

  ps_tasks = 0
  worker_replicas = 1
  worker_job_name = 'lonely_worker'
  task = 0
  is_chief = True
  master = ''

  if cluster_data and 'worker' in cluster_data:
    worker_replicas = len(cluster_data['worker']) + 1
  if cluster_data and 'ps' in cluster_data:
    ps_tasks = len(cluster_data['ps'])

  if worker_replicas > 1 and ps_tasks < 1:
    raise ValueError('At least 1 ps task is needed for distributed training.')

  if worker_replicas > = 1 and ps_tasks > 0:
    server = tf.train.Server(tf.train.ClusterSpec(cluster),protocol = 'grpc',
                             job_name = task_info.type, task_index = task_info.index)
    if task_info.type = = 'ps':
```

```
        server.join()
        return
    worker_job_name = '% s/task:% d' %(task_info.type,task_info.index)
    task = task_info.index
    is_chief = (task_info.type = = 'master')
    master = server.target
```

(5)训练主函数:记录训练日志,配置训练函数参数。

```
graph_rewriter_fn = None
  if 'graph_rewriter_config' in configs:
    graph_rewriter_fn = graph_rewriter_builder.build(
        configs['graph_rewriter_config'],is_training = True)

  trainer.train(
      create_input_dict_fn,
      model_fn,
      train_config,
      master,
      task,
      FLAGS.num_clones,
      worker_replicas,
      FLAGS.clone_on_cpu,
      ps_tasks,
      worker_job_name,
      is_chief,
      FLAGS.train_dir,
      graph_hook_fn = graph_rewriter_fn)
  print("模型训练完成!")

if __name__ = = '__main__':
  tf.app.run()
```

训练程序 train.py 的完整源代码文件参见 U5-train.py.pdf。

完整源代码

train.py

步骤4 训练模型

运行训练程序 train.py,如图 5.14 所示,程序读取配置文件 Vehicle.config 中定义的训练模型、训练参数、数据集,把训练日志和检查点保存到 checkpoint 目录中。

```
$ conda activate unit5
$ python train.py --logtostderr --train_dir checkpoint --pipeline_config_path data/vehicle.config
```

项目 5　智慧社区交通工具检测

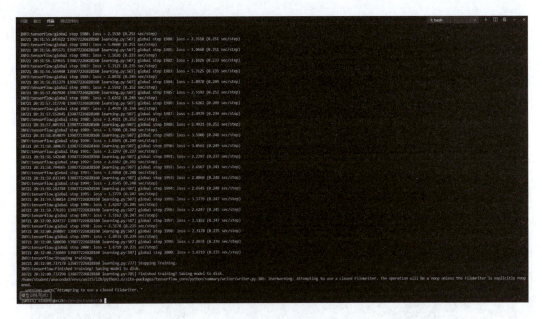

图 5.14　训练模型

步骤 5　可视化训练过程

在训练过程中打开 tensorboard 可以查看训练日志，如图 5.15 所示。训练日志中记录了模型分类损失、回归损失和总损失量的变化，通过 Losses 选项中的图表可以看到训练过程中的损失在不断变化，越到后面损失越小，说明模型对训练数据的拟合度越来越高。

＊注意需要把地址修改成对应的数据处理服务器地址，然后在浏览器中输入对应地址和端口号查看。

```
$ tensorboard --host 172.16.33.11 --port 8889 --logdir checkpoint/
```

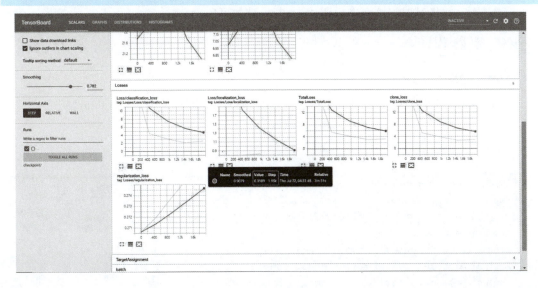

图 5.15　查看训练日志

步骤6 查看训练结果

进入 checkpoint 子目录，可以看到生成了多组模型文件，如图 5.16 所示。其中：
- model.ckpt-xxxx.meta 文件保存了计算图也就是神经网络的结构；
- model.ckpt-xxxx.data-xxxx 文件保存了模型的变量；
- model.ckpt-xxxx.index 文件保存了神经网络索引映射文件。

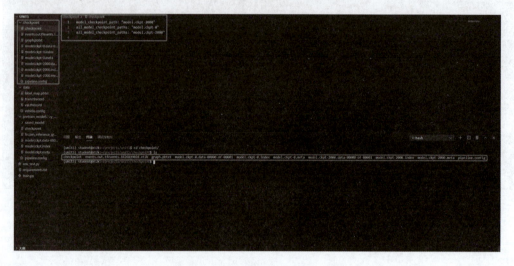

图 5.16 生成多组模型文件

任务 5 交通工具检测模型评估

任务描述

在本任务中，将对训练模型进行评估，判断模型的可用性。

任务操作

步骤1 创建评估文件

在开发环境中打开 /home/student/projects/unit5/ 目录，创建评估程序 eval.py。

(1) 导入模型的各个模块并定义输入参数。

```
import functools
import os
import tensorflow as tf
from object_detection.builders import dataset_builder
from object_detection.builders import graph_rewriter_builder
from object_detection.builders import model_builder
from object_detection.legacy import evaluator
from object_detection.utils import config_util
from object_detection.utils import label_map_util
```

```
os.environ["TF_CPP_MIN_LOG_LEVEL"] = '3'
tf.compat.v1.logging.set_verbosity(tf.compat.v1.logging.ERROR)
flags = tf.app.flags
flags.DEFINE_boolean('eval_training_data',False,'')
flags.DEFINE_string('checkpoint_dir','','')
flags.DEFINE_string('eval_dir','','Directory to write eval summaries.')
flags.DEFINE_string('pipeline_config_path','','Path to a pipeline config.')
flags.DEFINE_string('eval_config_path','','')
flags.DEFINE_string('input_config_path','','')
flags.DEFINE_string('model_config_path','','')
flags.DEFINE_boolean('run_once',False,'')
FLAGS = flags.FLAGS
```

(2) 评估主函数：加载模型配置。

```
@tf.contrib.framework.deprecated(None,'Use object_detection/model_main.py.')
def main(unused_argv):
  assert FLAGS.checkpoint_dir,''checkpoint_dir' is missing.'
  assert FLAGS.eval_dir,''eval_dir' is missing.'
  tf.gfile.MakeDirs(FLAGS.eval_dir)
  if FLAGS.pipeline_config_path:
    configs = config_util.get_configs_from_pipeline_file(
        FLAGS.pipeline_config_path)
    tf.gfile.Copy(
        FLAGS.pipeline_config_path,
        os.path.join(FLAGS.eval_dir,'pipeline.config'),
        overwrite = True)
  else:
    configs = config_util.get_configs_from_multiple_files(
        model_config_path = FLAGS.model_config_path,
        eval_config_path = FLAGS.eval_config_path,
        eval_input_config_path = FLAGS.input_config_path)
    for name,config in [('model.config',FLAGS.model_config_path),
                        ('eval.config',FLAGS.eval_config_path),
                        ('input.config',FLAGS.input_config_path)]:
      tf.gfile.Copy(config,os.path.join(FLAGS.eval_dir,name),overwrite = True)

  model_config = configs['model']
  eval_config = configs['eval_config']
  input_config = configs['eval_input_config']
  if FLAGS.eval_training_data:
    input_config = configs['train_input_config']

  model_fn = functools.partial(
      model_builder.build,model_config = model_config,is_training = False)
```

(3) 评估主函数：定义评估循环，并记录评估日志。

```
def get_next(config):
    return dataset_builder.make_initializable_iterator(
```

```
        dataset_builder.build(config)).get_next()

    create_input_dict_fn = functools.partial(get_next,input_config)

    categories = label_map_util.create_categories_from_labelmap(
        input_config.label_map_path)

    if FLAGS.run_once:
      eval_config.max_evals = 1

    graph_rewriter_fn = None
    if 'graph_rewriter_config' in configs:
      graph_rewriter_fn = graph_rewriter_builder.build(
          configs['graph_rewriter_config'],is_training = False)
```

(4)评估主函数:配置评估函数参数。

```
    evaluator.evaluate(
        create_input_dict_fn,
        model_fn,
        eval_config,
        categories,
        FLAGS.checkpoint_dir,
        FLAGS.eval_dir,
        graph_hook_fn = graph_rewriter_fn)
    print("模型评估完成!")

if __name__ == '__main__':
    tf.app.run()
```

评估程序 eval.py 的完整源代码文件参见 U5-eval.py.pdf。

完整源代码

eval.py

步骤2　评估已训练模型

运行评估程序 eval.py,如图 5.17 所示,程序读取配置文件 vehicle.config 中定义的训练模型、训练参数、数据集,读取 checkpoint 目录中的训练结果,把评估结果保存到 evaluation 目录中。

```
$ conda activate unit5
$ python eval.py --logtostderr --checkpoint_dir checkpoint --eval_dir evaluation --pipeline_config_path data/vehicle.config
```

在评估过程中,可以看到对不同类别的评估结果。

项目 5　智慧社区交通工具检测

图 5.17　评估模型

步骤 3　查看评估结果

利用 tensorboard 工具查看评估结果，如图 5.18 所示。

＊注意需要把地址修改成对应的数据处理服务器地址，然后在浏览器中输入对应地址和端口号查看。

```
$ tensorboard --host 172.16.33.11 --port 8889 --logdir evaluation/
```

步骤 4　分析模型可用性

在浏览器中查看各类别的平均精确度（AP）值，越接近 1 说明模型的可用性越高。此时图上显示 step 为 2k，说明该模型是训练到 2 000 步时保存下来的，对应 model.ckpt-2000 训练模型。

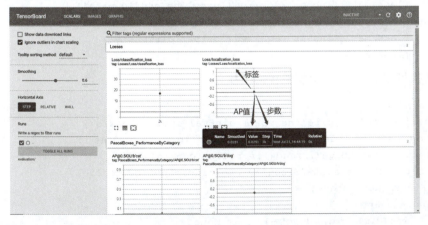

图 5.18　查看评估结果

任务6 交通工具检测模型测试

任务描述

在本任务中,将把已经评估为可用性较强的模型,导出成为可测试的冻结图模型,用测试数据进行测试。

任务操作

步骤1 创建导出文件

在开发环境中打开/home/student/projects/unit5/目录,创建导出程序 export_fz.py。

(1)导入模型转换模块,定义输入参数。

```python
import os
import tensorflow as tf
from google.protobuf import text_format
from object_detection import exporter
from object_detection.protos import pipeline_pb2

os.environ["TF_CPP_MIN_LOG_LEVEL"] = '3'
tf.compat.v1.logging.set_verbosity(tf.compat.v1.logging.ERROR)
slim = tf.contrib.slim
flags = tf.app.flags

flags.DEFINE_string('input_type','image_tensor','')
flags.DEFINE_string('input_shape',None,'[None,None,None,3]')
flags.DEFINE_string('pipeline_config_path',None,'Path to a pipeline config.')
flags.DEFINE_string('trained_checkpoint_prefix',None,'path/to/model.ckpt')
flags.DEFINE_string('output_directory',None,'Path to write outputs.')
flags.DEFINE_string('config_override','','')
flags.DEFINE_boolean('write_inference_graph',False,'')
tf.app.flags.mark_flag_as_required('pipeline_config_path')
tf.app.flags.mark_flag_as_required('trained_checkpoint_prefix')
tf.app.flags.mark_flag_as_required('output_directory')
FLAGS = flags.FLAGS
```

(2)转换模型主函数,调用模型转换函数。

```python
def main(_):
  pipeline_config = pipeline_pb2.TrainEvalPipelineConfig()
  with tf.gfile.GFile(FLAGS.pipeline_config_path,'r') as f:
    text_format.Merge(f.read(),pipeline_config)
  text_format.Merge(FLAGS.config_override,pipeline_config)
  if FLAGS.input_shape:
    input_shape = [
        int(dim) if dim != '-1' else None
        for dim in FLAGS.input_shape.split(',')
```

```
    ]
  else:
    input_shape = None
  exporter.export_inference_graph(
      FLAGS.input_type,pipeline_config,FLAGS.trained_checkpoint_prefix,
      FLAGS.output_directory,input_shape = input_shape,
      write_inference_graph = FLAGS.write_inference_graph)
  print("模型转换完成!")

if __name__ = = '__main__':
  tf.app.run()
```

导出程序 export_fz.py 的完整源代码文件参见 U5-export_fz.py.pdf。

完整源代码

export_fz.py

步骤2 导出冻结图模型

运行导出程序 export_fz.py,如图 5.19 所示,程序将读取配置文件 Vehicle.config 中定义的配置,读取 checkpoint 目录中的 model.ckpt-2000 训练模型,导出为冻结图模型,并保存到 frozen_models 目录中。

```
$ conda activate unit5
$ python export_fz.py --input_type image_tensor --pipeline_config_path data/vehicle.config --trained_checkpoint_prefix checkpoint/model.ckpt-2000 --output_directory frozen_models
```

图 5.19 导出冻结图模型

步骤3 创建测试文件

在开发环境中打开/home/student/projects/unit5/目录,创建测试文件 detect.py。

(1) 导入测试所需模块和可视化函数,定义输入参数。

```python
import numpy as np
import os
import tensorflow as tf
import matplotlib.pyplot as plt
from PIL import Image
from object_detection.utils import label_map_util
from object_detection.utils import visualization_utils as vis_util
from object_detection.utils import ops as utils_ops

os.environ["TF_CPP_MIN_LOG_LEVEL"] = '3'
tf.compat.v1.logging.set_verbosity(tf.compat.v1.logging.ERROR)
detect_img = '/home/student/data/vehicle/test/1.jpg'
result_img = '/home/student/projects/unit5/img/1_result.jpg'
MODEL_NAME = 'frozen_models'
PATH_TO_FROZEN_GRAPH = MODEL_NAME + '/frozen_inference_graph.pb'
PATH_TO_LABELS = 'data/label_map.pbtxt'
```

(2) 加载模型计算图和数据标签。

```python
detection_graph = tf.Graph()
with detection_graph.as_default():
    od_graph_def = tf.compat.v1.GraphDef()
    with tf.io.gfile.GFile(PATH_TO_FROZEN_GRAPH,'rb') as fid:
        serialized_graph = fid.read()
        od_graph_def.ParseFromString(serialized_graph)
        tf.import_graph_def(od_graph_def,name='')
category_index = label_map_util.create_category_index_from_labelmap(
        PATH_TO_LABELS,use_display_name = True)
```

(3) 图片数据转换函数。

```python
def load_image_into_numpy_array(image):
    (im_width,im_height) = image.size
    return np.array(image.getdata()).reshape((im_height,im_width,3)).astype(np.uint8)
```

(4) 单张图像检测函数。

```python
def run_inference_for_single_image(image,graph):
    with graph.as_default():
        with tf.compat.v1.Session() as sess:
            ops = tf.compat.v1.get_default_graph().get_operations()
            all_tensor_names = {output.name for op in ops for output in op.outputs}
            tensor_dict = {}
            for key in ['num_detections','detection_boxes','detection_scores',
                'detection_classes','detection_masks']:
                tensor_name = key + ':0'
                if tensor_name in all_tensor_names:
```

```python
            tensor_dict[key] = tf.compat.v1.get_default_graph().get_tensor_by_name(tensor_name)
        if 'detection_masks' in tensor_dict:
            detection_boxes = tf.squeeze(tensor_dict['detection_boxes'],[0])
            detection_masks = tf.squeeze(tensor_dict['detection_masks'],[0])
            real_num_detection = tf.cast(tensor_dict['num_detections'][0],tf.int32)
            detection_boxes = tf.slice(detection_boxes,[0,0],[real_num_detection,-1])
            detection_masks = tf.slice(detection_masks,[0,0,0],[real_num_detection,-1,-1])
            detection_masks_reframed = utils_ops.reframe_box_masks_to_image_masks(
                detection_masks,detection_boxes,image.shape[1],image.shape[2])
            detection_masks_reframed = tf.cast(tf.greater(detection_masks_reframed,
                0.5),tf.uint8)
            tensor_dict['detection_masks'] = tf.expand_dims(detection_masks_reframed,0)
            image_tensor = tf.compat.v1.get_default_graph().get_tensor_by_name('image_tensor:0')

        output_dict = sess.run(tensor_dict,feed_dict = {image_tensor:image})

        output_dict['num_detections'] = int(output_dict['num_detections'][0])
        output_dict['detection_classes'] = output_dict['detection_classes'][0].astype(np.int64)
        output_dict['detection_boxes'] = output_dict['detection_boxes'][0]
        output_dict['detection_scores'] = output_dict['detection_scores'][0]
        if 'detection_masks' in output_dict:
            output_dict['detection_masks'] = output_dict['detection_masks'][0]
    return output_dict
```

（5）输入图片数据，检测输入数据，保存检测结果图。

```python
image = Image.open(detect_img)
image_np = load_image_into_numpy_array(image)
# 转化输入图片为 shape = [1,None,None,3]
image_np_expanded = np.expand_dims(image_np,axis = 0)
output_dict = run_inference_for_single_image(image_np_expanded,detection_graph)
vis_util.visualize_boxes_and_labels_on_image_array(
    image_np,
    output_dict['detection_boxes'],
    output_dict['detection_classes'],
    output_dict['detection_scores'],
    category_index,
    instance_masks = output_dict.get('detection_masks'),
    use_normalized_coordinates = True,
    line_thickness = 6)
plt.figure()
plt.axis('off')
plt.imshow(image_np)
plt.savefig(result_img,bbox_inches = 'tight',pad_inches = 0)
print("测试% s 完成,结果保存在% s" % (detect_img,result_img))
```

测试程序 detect.py 的完整源代码文件参见 U5-detect.py.pdf。

完整源代码

detect.py

步骤 4　测试并查看结果

创建 img 目录存放测试结果，运行测试程序 detect.py，如图 5.20 所示，并查看输出的结果图片，如图 5.21 所示。

```
$ mkdir img
$ python detect.py
```

图 5.20　测试模型

图 5.21　查看测试结果

任务 7　交通工具检测模型部署

任务描述

在本任务中，将把经过测试确认可用的模型，转换成为应用部署支持的格式，然后将模型文件部署到边缘计算设备上，并实现人工智能应用的集成。

任务操作

步骤1　创建导出程序

在开发环境中打开/home/student/projects/unit5/目录,创建导出程序export_pb.py。

(1)导入模型转换模块,定义输入参数。

```python
import os
import tensorflow as tf
from google.protobuf import text_format
from object_detection import export_tflite_ssd_graph_lib
from object_detection.protos import pipeline_pb2

os.environ["TF_CPP_MIN_LOG_LEVEL"] = '3'
tf.compat.v1.logging.set_verbosity(tf.compat.v1.logging.ERROR)
flags = tf.app.flags
flags.DEFINE_string('output_directory',None,'Path to write outputs.')
flags.DEFINE_string('pipeline_config_path',None,'')
flags.DEFINE_string('trained_checkpoint_prefix',None,'Checkpoint prefix.')
flags.DEFINE_integer('max_detections',10,'')
flags.DEFINE_integer('max_classes_per_detection',1,'')
flags.DEFINE_integer('detections_per_class',100,'')
flags.DEFINE_bool('add_postprocessing_op',True,'')
flags.DEFINE_bool('use_regular_nms',False,'')
flags.DEFINE_string('config_override','','')
FLAGS = flags.FLAGS
```

(2)调用模型转换函数,完成模型转换。

```python
def main(argv):
  flags.mark_flag_as_required('output_directory')
  flags.mark_flag_as_required('pipeline_config_path')
  flags.mark_flag_as_required('trained_checkpoint_prefix')

  pipeline_config = pipeline_pb2.TrainEvalPipelineConfig()

  with tf.gfile.GFile(FLAGS.pipeline_config_path,'r') as f:
    text_format.Merge(f.read(),pipeline_config)
  text_format.Merge(FLAGS.config_override,pipeline_config)
  export_tflite_ssd_graph_lib.export_tflite_graph(
      pipeline_config,FLAGS.trained_checkpoint_prefix,FLAGS.output_directory,
      FLAGS.add_postprocessing_op,FLAGS.max_detections,
      FLAGS.max_classes_per_detection,FLAGS.use_regular_nms)
  print("模型转换完成!")

if __name__ == '__main__':
  tf.app.run(main)
```

导出程序export_pb.py的完整源代码文件参见U5-export_pb.py.pdf。

export_pb.py

步骤 2　导出 pb 文件

运行导出程序 export_pb.py，读取配置文件 vehicle.config 中定义的参数，读取 checkpoint 目录中的训练结果，把 tflite_pb 模型图保存到 tflite_models 目录中。

```
$ conda activate unit5
$ python export_pb.py --pipeline_config_path data/vehicle.config --trained_checkpoint_prefix checkpoint/model.ckpt-2000 --output_directory tflite_models
```

步骤 3　创建转换程序

在开发环境中打开 /home/student/projects/unit5/ 目录，创建转换程序 pb_to_tflite.py。

（1）导入模块，定义输入参数。

```python
import os
import tensorflow as tf

os.environ["TF_CPP_MIN_LOG_LEVEL"] = '3'
tf.compat.v1.logging.set_verbosity(tf.compat.v1.logging.ERROR)
flags = tf.app.flags
flags.DEFINE_string('pb_path','tflite_models/tflite_graph.pb','tflite pb file.')
flags.DEFINE_string('tflite_path','tflite_models/zy_ssd.tflite','output tflite.')
FLAGS = flags.FLAGS
```

（2）转换为 tflite 模型。

```python
def convert_pb_to_tflite(pb_path,tflite_path):
    # 模型输入节点
    input_tensor_name = ["normalized_input_image_tensor"]
    input_tensor_shape = {"normalized_input_image_tensor":[1,300,300,3]}
    # 模型输出节点
    classes_tensor_name = ['TFLite_Detection_PostProcess','TFLite_Detection_PostProcess:1','TFLite_Detection_PostProcess:2','TFLite_Detection_PostProcess:3']
    # 转换为 tflite 模型
    converter = tf.lite.TFLiteConverter.from_frozen_graph(pb_path,
                                                          input_tensor_name,
                                                          classes_tensor_name,
                                                          input_tensor_shape)
    converter.allow_custom_ops = True
    converter.optimizations = [tf.lite.Optimize.DEFAULT]
    tflite_model = converter.convert()
```

（3）tflite 模型写入。

```
converter.allow_custom_ops = True
    converter.optimizations = [tf.lite.Optimize.DEFAULT]
    tflite_model = converter.convert()
    # 模型写入
    if not tf.gfile.Exists(os.path.dirname(tflite_path)):
        tf.gfile.MakeDirs(os.path.dirname(tflite_path))
    with open(tflite_path,"wb") as f:
        f.write(tflite_model)
    print("Save tflite model at % s" % tflite_path)
    print("模型转换完成!")

if __name__ == '__main__':
    convert_pb_to_tflite(FLAGS.pb_path, FLAGS.tflite_path)
```

转换程序 pb_to_tflite.py 的完整源代码文件参见 U5-pb_to_tflite.py.pdf。

pb_to_tflite.py

步骤 4　转换 tflite 文件

运行文件程序 pb_to_tflite.py。

```
$ python pb_to_tflite.py
```

步骤 5　创建推理执行程序

在开发环境中打开 /home/student/projects/unit5/tflite_models 目录，创建推理执行程序 func_detection_img.py。

（1）导入所用模块。

```
import os
import cv2
import numpy as np
import sys
import glob
import importlib.util
import base64
```

（2）定义模型和数据推理器。

```
def update_image(image_data, GRAPH_NAME = 'zy_ssd.tflite', min_conf_threshold = 0.5, use_TPU = False, model_dir = 'util'):
    from tflite_runtime.interpreter import Interpreter
    CWD_PATH = os.getcwd()
```

```
            PATH_TO_CKPT = os.path.join(CWD_PATH,model_dir,GRAPH_NAME)

            labels = ['bicycle','car','motorbike','bus']

            interpreter = Interpreter(model_path = PATH_TO_CKPT)

            interpreter.allocate_tensors()

            input_details = interpreter.get_input_details()
            output_details = interpreter.get_output_details()
            height = input_details[0]['shape'][1]
            width = input_details[0]['shape'][2]

            floating_model = (input_details[0]['dtype'] = = np.float32)

            input_mean = 127.5
            input_std = 127.5
```

(3)输入图像并转换图像数据为张量。

```
# base64 解码
img_data = base64.b64decode(image_data)
# 转换为 np 数组
img_array = np.fromstring(img_data,np.uint8)
# 转换成 opencv 可用格式
image = cv2.imdecode(img_array,cv2.COLOR_RGB2BGR)

image_rgb = cv2.cvtColor(image,cv2.COLOR_BGR2RGB)
imH,imW,_ = image.shape
image_resized = cv2.resize(image_rgb,(width,height))
input_data = np.expand_dims(image_resized,axis = 0)

if floating_model:
    input_data = (np.float32(input_data)-input_mean)/input_std

interpreter.set_tensor(input_details[0]['index'],input_data)
interpreter.invoke()

boxes = interpreter.get_tensor(output_details[0]['index'])[0]
classes = interpreter.get_tensor(output_details[1]['index'])[0]
scores = interpreter.get_tensor(output_details[2]['index'])[0]
```

(4)检测图片,并可视化输出结果。

```
for i in range(len(scores)):
    if((scores[i] >min_conf_threshold) and(scores[i] < =1.0)):
        ymin = int(max(1,(boxes[i][0] * imH)))
        xmin = int(max(1,(boxes[i][1] * imW)))
        ymax = int(min(imH,(boxes[i][2] * imH)))
        xmax = int(min(imW,(boxes[i][3] * imW)))
```

```
        cv2.rectangle(image,(xmin,ymin),(xmax,ymax),(10,255,0),2)

        object_name = labels[int(classes[i])]
        label = '% s:% d% % ' %(object_name,int(scores[i] * 100))
        labelSize,baseLine = cv2.getTextSize(label,cv2.FONT_HERSHEY_SIMPLEX,0.7,2)
        label_ymin = max(ymin,labelSize[1] +10)
        cv2.rectangle(image,(xmin,label_ymin-labelSize[1]-10),
                    (xmin + labelSize[0],label_ymin + baseLine-10),(255,255,255),
cv2.FILLED)
        cv2.putText(image,label,(xmin,label_ymin-7),cv2.FONT_HERSHEY_SIMPLEX,0.7,
(0,0,0),2)

        image_bytes = cv2.imencode('.jpg',image)[1].tostring()
        image_base64 = base64.b64encode(image_bytes).decode()
        return image_base64
```

推理执行程序 func_detection_img.py 的完整源代码文件参见 U5-func_detection_img.py.pdf。

完整源代码

func_detection_img.py

步骤6　部署到边缘设备

把模型 zy_ssd.tflite 文件、推理执行程序 func_detection_img.py 文件复制到边缘计算设备中。

* 注意把 IP 地址修改成对应的推理机地址。

```
    $ scp tflite_models/zy_ssd.tflite student@172.16.33.118:/home/student/zy-panel-check/util/
    $ scp tflite_models/func_detection_img.py student@172.16.33.118:/home/student/zy-panel-check/util/
```

通过平台上的"应用部署"按钮,上传或输入图片 URL 检测,应用部署成功结果如图5.22所示。

图 5.22　应用部署成功

人工智能图像识别项目实践

项目小结

通过亲自动手完成数据标注、模型训练、模型导出等任务,你实现了一个交通工具的机器识别模型,并部署到边缘计算设备上,通过了模型测试。同时,通过与另外一组同事联调,实现了"车不入位"及"违章停车"的报警,圆满完成了项目经理交给的工作。项目交付之后,小区业主反馈"现在开车上下班比以前通畅多了,违停的车辆也少了很多,还经常看到街道办的巡防队员做交通疏导"。

多学一点:深度学习在机器学习中起着重要的作用,它通过建立、模拟人脑的分层结构来实现对外部输入的数据进行从低级到高级的特征提取,从而能够解释外部数据。深度学习是一个多层次的学习,逐层学习并把学习的知识传递给下一层,通过这种方式,就可以实现对输入信息进行分级表达。深度学习的实质就是通过建立、模拟人脑的分层结构,对外部输入的声音、图像、文本等数据进行从低级到高级的特征提取,从而能够解释外部数据。与传统学习结构相比,深度学习更加强调模型结构的深度,通常含有多层的隐层节点,而且在深度学习中,特征学习至关重要,通过特征的逐层变换完成最后的预测和识别。祝愿你在未来的学习中掌握更多的技能,在实际工作中灵活运用,成为一名优秀的工程师。

项目 6

节能洗车房车牌识别

项目背景

学电子信息的你加入了一家节能环保企业,公司的主营产品是节能型洗车房。由于节水节电而且可自动洗车,产品迅速得到了市场和资本的认可。公司决定继续投入研发新一代产品:在节能洗车房的基础上实现无人值守的功能。新产品需要通过图像识别检测出车牌号码,车主通过扫码支付后,洗车房的卷帘门自动开启。项目启动会上,领导告诉大家:无人值守是新一代节能洗车房大规模推广的利器,可以帮助公司把传统的直营模式变成直营与加盟共存的经营模式,能够极大降低加盟门槛、快速扩大市场份额,对规范洗车行业市场、降低城市碳排放量有重要意义,所以请大家务必重视这个项目的实用性和通用性。新产品的研发由公司总工亲自挂帅,他对团队寄予厚望,作为人工智能训练师,你被分在图像识别团队。项目经理为你提供了数百张原始车牌图片,并配备了一位资深算法工程师为你提供预训练模型,要求你在硬件设计定稿打样之前,完成车牌识别的模型训练,能够识别京牌车号,并部署到边缘计算设备上测试通过。为了给后续实际上线工作提供可靠的基础,你的工作需要在一周内完成,请尽快开始。

提示:对车牌识别应用来说,重新设计一个全新的深度学习神经网络模型,然后用数以万计的图片数据区训练,对公司来说实现的难度和成本非常高。因此基于深度学习预训练模型,用百张数量级的图片做迁移学习,是一个容易实现的技术路径。经过分析,梳理出的需求是通过图片识别车牌、车牌所属地汉字、车牌字母、阿拉伯数字。由于只有一周时间作为概念验证,因此车牌所属地汉字只需要识别"京",字母没有 I 和 O,所以一共需要建立 36 个类别。

任务 1　数据准备

任务描述

在本任务中,将获得项目团队提供的车牌图片数据集,并把这些图片数据导入人工智能数据处理平台中,分别保存在训练、验证、测试等不同的目录中。

 任务操作

步骤 1 数据采集

项目经理已经准备好了相关图片数据。

步骤 2 数据整理

把数据下载到工作目录,解压缩。在终端命令行窗口中执行以下操作。

* 注意第二行命令需要把地址修改成对应的资源平台地址。

```
$ cd ~/data
$ wget http://172.16.33.72/dataset/plate.tar.gz
$ tar zxvf plate.tar.gz
```

在终端命令行窗口中执行以下操作,查看解压缩后的目录,如图 6.1 所示。

```
$ cd ~/data/plate
$ ls
```

图 6.1 查看解压缩后的目录

任务 2 工程环境准备

 任务描述

在本任务中,将创建本项目的开发环境,并做基础配置,满足数据标注、模型训练、评估和部署的基础条件。

 任务操作

步骤 1 创建工程目录

在开发环境中打开为本项目创建的工程目录,在终端命令行窗口中执行以下操作:

```
$ mkdir ~/projects/unit6
$ mkdir ~/projects/unit6/data
$ cd ~/projects/unit6
```

步骤 2 创建开发环境

在 Python 3.6 中创建名为 unit6 的虚拟环境。

```
$ conda create -n unit6 python=3.6
```

输入 y 继续完成,然后执行以下操作激活开发环境。

```
$ conda activate unit6
```

步骤3　配置GPU环境

安装 tensorflow-gpu 1.15 环境。

```
$ conda install tensorflow-gpu=1.15
```

输入 y 继续完成 GPU 环境的配置,如图 6.2 所示。

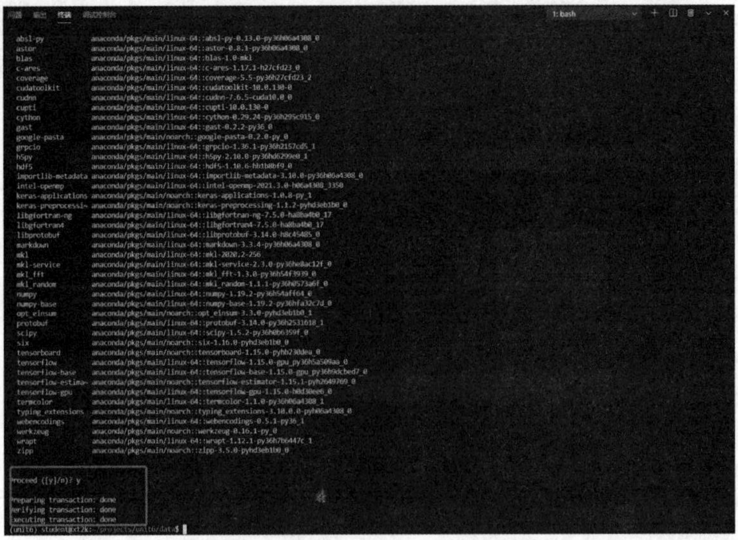

图 6.2　完成 GPU 环境的配置

步骤4　配置依赖环境

(1) 在开发环境中打开/home/student/projects/unit6 目录,创建依赖清单文件 requirements.txt。

(2) 把以下内容写到 requirements.txt 清单文件中,然后执行命令,安装依赖库环境,完成依赖环境的配置如图 6.3 所示。

```
# requirements.txt
Cython
contextlib2
matplotlib
pillow
lxml
jupyter
pycocotools
click
PyYAML
joblib
autopep8

#执行以下命令
$ conda activate unit6
$ pip install -r requirements.txt
```

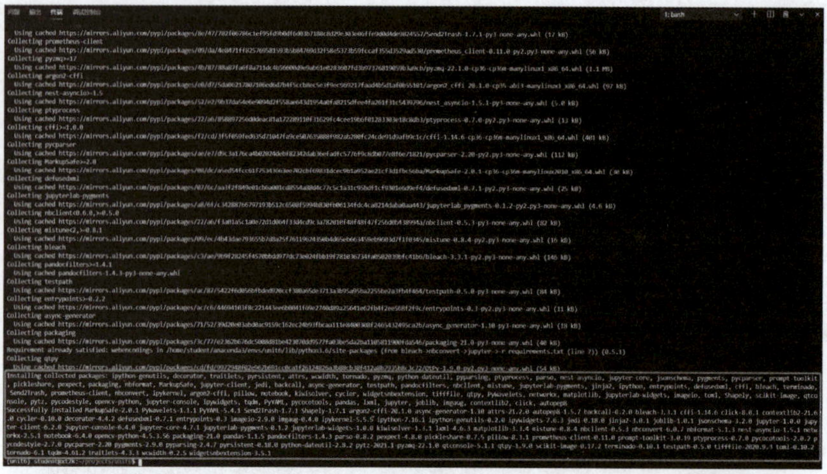

图 6.3　完成依赖环境的配置

步骤 5　配置图像识别库环境

（1）安装中育 object_detection 库和中育 slim 库。
（2）在终端命令行窗口中执行以下操作，完成后删除安装程序。
＊注意第一行和第三行命令需要把地址修改成对应的资源平台地址。

```
$ wget http://172.16.33.72/dataset/dist/zy_od_1.0.tar.gz
$ pip install zy_od_1.0.tar.gz
$ wget http://172.16.33.72/dataset/dist/zy_slim_1.0.tar.gz
$ pip install zy_slim_1.0.tar.gz
$ rm zy_od_1.0.tar.gz zy_slim_1.0.tar.gz
```

步骤 6　验证环境

在终端命令行窗口中执行以下操作，查看验证结果，如图 6.4 所示。

```
$ wget http://172.16.33.72/dataset/script/env_test.py
$ python env_test.py
```

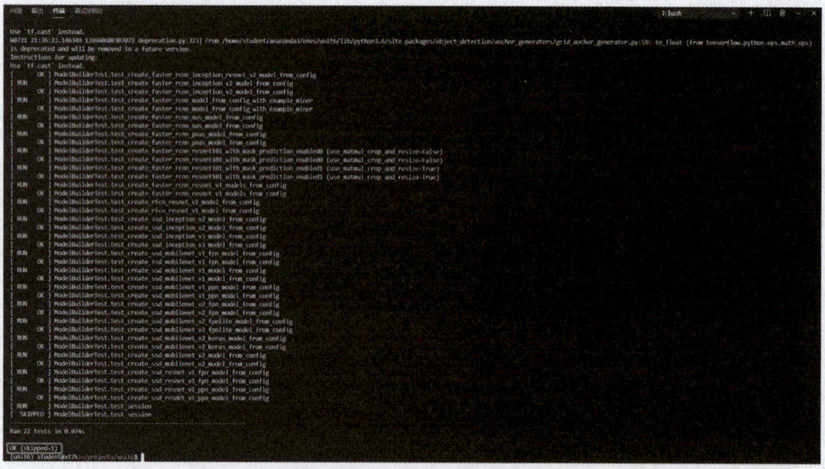

图 6.4　验证结果

项目 6 节能洗车房车牌识别

任务 3　车牌图片数据标注

任务描述

在本任务中,将使用图片标注工具完成数据标注,导出为数据集文件,并保存标签映射文件。

任务操作

步骤 1　添加标注标签

(1)创建名称为"节能洗车房车牌识别"标注项目,如图 6.5 所示。

(2)添加 36 类标签,分别为 plate,jing,A,B,C,D,E,F,G,H,J,K,L,M,N,P,Q,R,S,T,U,V,W,X,Y,Z,0,1,2,3,4,5,6,7,8,9。

(3)注意设置为不同的颜色标签以示区分。

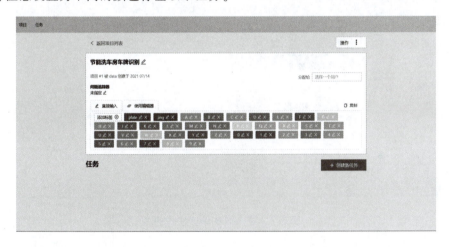

图 6.5　创建标注项目

步骤 2　创建训练集任务

(1)任务名称为"节能洗车房车牌识别训练集"。

(2)任务子集选择 Train。

(3)选择文件使用"连接共享文件",选中本项目任务 1 中整理的 train 子目录,如图 6.6 所示。

(4)选择提交,完成创建。

步骤 3　标注训练集数据

(1)打开"节能洗车房车牌识别训练集",单击左下方的"作业#"进入标注作业,如图 6.7 所示。

(2)使用加锁,可以避免对已标注对象的误操作;将一张图片中的对象标注完成后,单击上方工具栏中的"下一帧"按钮继续标注,如图 6.8 所示,直至整个数据集标注完成。

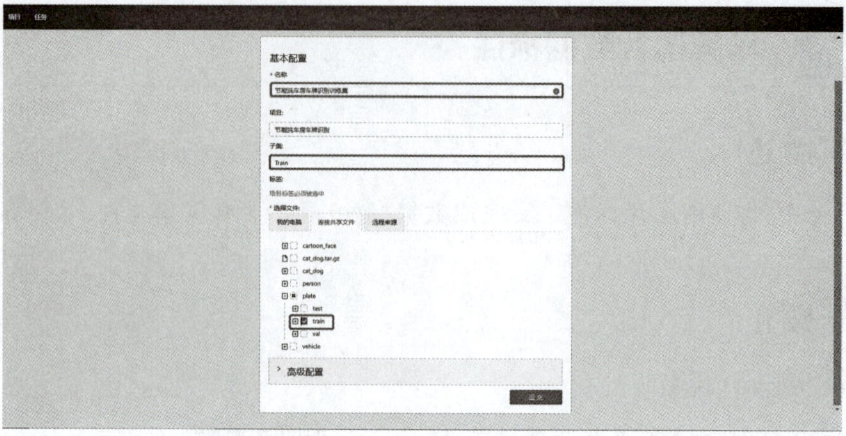

图6.6 创建训练任务

图6.7 进入训练集标注作业

图6.8 标注图片

步骤4 导出标注训练集

(1)选择"菜单"→"导出为数据集"→"TFRecord 1.0"命令,如图6.9所示。

图6.9 导出为数据集

(2)标注完成的数据导出后是一个压缩包 zip 文件,保存在浏览器默认的下载路径中;将该文件解压缩,并把 default.tfrecord 重命名为 train.tfrecord。

步骤5 创建验证集任务

(1)任务名称为"节能洗车房车牌识别验证集"。

(2)任务子集选择 Validation。

(3)选择文件使用"连接共享文件",选中本项目任务1中整理的 val 子目录,如图6.10所示。

(4)单击"提交"按钮,完成创建。

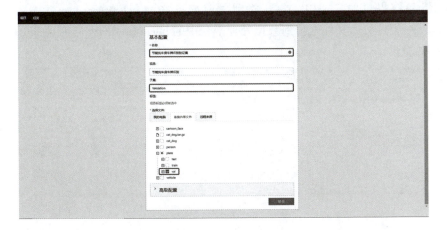

图6.10 创建验证任务

步骤6 标注验证集数据

(1)打开"节能洗车房车牌识别验证集",单击左下方的"作业#"进入标注作业,如图6.11所示。

图 6.11　进入验证集标注作业

（2）将一张图片中的对象标注完成后，单击上方工具栏中的"下一帧"按钮继续标注，直至整个数据集标注完成。

步骤 7　导出标注验证集

（1）选择"菜单"→"导出为数据集"→"TFRecord 1.0"命令。

（2）标注完成的数据集导出后是一个压缩文件，保存在浏览器默认的下载路径中。

（3）将该文件解压缩后会得到 default.tfrecord 和 label_map.pbtxt 文件，把 default.tfrecord 重命名为 val.tfrecord。

（4）找到之前保存好的 train.tfrecord 文件，并将 val.tfrecord、train.tfrecord、label_map.pbtxt 三个文件存放到一起备用。

步骤 8　上传文件

（1）打开系统提供的 winSCP 工具，找到之前准备好的 val.tfrecord、train.tfrecord、label_map.pbtxt 文件，把这三个文件上传到数据处理平台中的 home/student/projects/unit3/data 目录下，如图 3.12 所示。

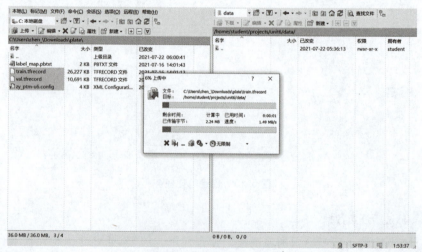

图 6.12　上传标注数据文件

（2）上传成功后，在平台上可以看到以下三个文件，如图 6.13 所示。

图 6.13　数据标注文件上传成功

任务 4　车牌识别模型训练

任务描述

在本任务中，将搭建训练模型、配置预训练模型参数，对已标注的数据集进行训练，得到训练模型，并学习使用可视化的工具查看训练结果。

任务操作

步骤 1　搭建模型

在开发环境中打开准备预训练模型相关目录。

```
$ cd ~/projects/unit6
$ mkdir pretrain_models
$ cd pretrain_models
```

下载算法团队提供的预训练模型，并解压缩。

＊注意需要把地址修改成对应的资源平台地址。

```
$ wget http://172.16.33.72/dataset/dist/zy_ptm_u6.tar.gz
$ tar zxvf zy_ptm_u6.tar.gz
$ rm zy_ptm_u6.tar.gz
```

步骤 2　配置训练模型

在开发环境中打开 /home/student/projects/unit6/data 目录，创建训练模型配置文件 plate.config。

（1）主干网络配置。主干网络是整个模型训练的基础，标记了当前模型识别的物体类别等重要信息。本项目车牌识别为 36 类，因此 num_classes 为 36。

```
num_classes:36
box_coder {
  faster_rcnn_box_coder {
    y_scale:10.0
    x_scale:10.0
    height_scale:5.0
    width_scale:5.0
  }
}
matcher {
  argmax_matcher {
```

```
      matched_threshold:0.5
      unmatched_threshold:0.5
      ignore_thresholds:false
      negatives_lower_than_unmatched:true
      force_match_for_each_row:true
    }
  }
  similarity_calculator {
    iou_similarity {
    }
  }
```

（2）先验框配置和图片分辨率配置。image_resizer 表示模型输入图片分辨率，此处为标准的 320×640，因此 height 为 320，width 为 640。

```
  anchor_generator {
    ssd_anchor_generator {
      num_layers:6
      min_scale:0.2
      max_scale:0.95
      aspect_ratios:1.0
      aspect_ratios:2.0
      aspect_ratios:0.5
      aspect_ratios:3.0
      aspect_ratios:0.3333
    }
  }
  image_resizer {
    fixed_shape_resizer {
      height:320
      width:640
    }
  }
```

（3）边界预测框配置。

```
  box_predictor {
    convolutional_box_predictor {
      min_depth:0
      max_depth:0
      num_layers_before_predictor:0
      use_dropout:false
      dropout_keep_probability:0.8
      kernel_size:1
      box_code_size:4
      apply_sigmoid_to_scores:false
      conv_hyperparams {
        activation:RELU_6,
        regularizer {
          l2_regularizer {
```

```
        weight:0.00004
      }
    }
    initializer {
      truncated_normal_initializer {
        stddev:0.03
        mean:0.0
      }
    }
    batch_norm {
      train:true,
      scale:true,
      center:true,
      decay:0.9997,
      epsilon:0.001,
    }
  }
 }
}
```

(4) 特征提取网络配置。

```
feature_extractor {
  type:'ssd_mobilenet_v2'
  min_depth:16
  depth_multiplier:1.0
  use_explicit_padding:true
  conv_hyperparams {
    activation:RELU_6,
    regularizer {
      l2_regularizer {
        weight:0.00004
      }
    }
    initializer {
      truncated_normal_initializer {
        stddev:0.03
        mean:0.0
      }
    }
    batch_norm {
      train:true,
      scale:true,
      center:true,
      decay:0.9997,
      epsilon:0.001,
    }
  }
}
```

(5) 模型损失函数配置。

```
loss {
  classification_loss {
    weighted_sigmoid {
    }
  }
  localization_loss {
    weighted_smooth_l1 {
    }
  }
  hard_example_miner {
    num_hard_examples:3000
    iou_threshold:0.99
    loss_type:CLASSIFICATION
    max_negatives_per_positive:3
    min_negatives_per_image:3
  }
  classification_weight:1.0
  localization_weight:1.0
}
normalize_loss_by_num_matches:true
post_processing {
  batch_non_max_suppression {
    score_threshold:1e-8
    iou_threshold:0.6
    max_detections_per_class:100
    max_total_detections:100
  }
  score_converter:SIGMOID
}
```

(6) 训练集数据配置。batch_size 代表批处理每次迭代的数据量，initial_learning_rate 代表初始学习率，fine_tune_checkpoint 指向预训练模型文件，input_path 指向训练集的 tfrecord 文件，label_map_path 指向标签映射文件。

```
train_config:{
  batch_size:16
  sync_replicas:true
  startup_delay_steps:0
  replicas_to_aggregate:4
  optimizer {
    rms_prop_optimizer:{
      learning_rate:{
        cosine_decay_learning_rate {
          learning_rate_base:.02
          total_steps:50000
          warmup_learning_rate:.002
          warmup_steps:2000
        }
      }
```

```
      momentum_optimizer_value:0.9
      decay:0.9
      epsilon:1.0
    }
  }
  fine_tune_checkpoint:"pretrain_models/zy_ptm_u6/model.ckpt"
  fine_tune_checkpoint_type:  "detection"
  num_steps:2000
  data_augmentation_options {
    random_horizontal_flip {
    }
  }
  data_augmentation_options {
    ssd_random_crop_fixed_aspect_ratio {
    }
  }
}

train_input_reader:{
  tf_record_input_reader {
    input_path:"data/train.tfrecord"
  }
  label_map_path:"data/label_map.pbtxt"
}
```

（7）验证集数据配置。num_examples 代表验证集样本数量，input_path 指向验证集的 tfrecord 文件，label_map_path 指向标签映射文件。

```
eval_config:{
  num_examples:48
  max_evals:1
}

eval_input_reader:{
  tf_record_input_reader {
    input_path:"data/val.tfrecord"
  }
  label_map_path:"data/label_map.pbtxt"
  shuffle:false
  num_readers:1
}
```

模型配置文件 plate.config 的完整源代码文件参见 U6-plate.config.pdf。

plate.config

步骤3 创建训练程序

在开发环境中打开/home/student/projects/unit6/目录,创建训练程序 train.py。

(1)导入训练所需模块和函数。

```python
import functools
import json
import os
import tensorflow as tf
from object_detection.builders import dataset_builder
from object_detection.builders import graph_rewriter_builder
from object_detection.builders import model_builder
from object_detection.legacy import trainer
from object_detection.utils import config_util
```

(2)定义输入参数。

```python
os.environ["TF_CPP_MIN_LOG_LEVEL"] = '3'
tf.logging.set_verbosity(tf.logging.INFO)
flags = tf.app.flags
flags.DEFINE_string('master','','')
flags.DEFINE_integer('task',0,'task id')
flags.DEFINE_integer('num_clones',1,'')
flags.DEFINE_boolean('clone_on_cpu',False,'')
flags.DEFINE_integer('worker_replicas',1,'')
flags.DEFINE_integer('ps_tasks',0,'')
flags.DEFINE_string('train_dir','','Directory to save the checkpoints and training summaries.')
flags.DEFINE_string('pipeline_config_path','','Path to a pipeline config.')
flags.DEFINE_string('train_config_path','','Path to a train_pb2.TrainConfig.')
flags.DEFINE_string('input_config_path','','Path to an input_reader_pb2.InputReader.')
flags.DEFINE_string('model_config_path','','Path to a model_pb2.DetectionModel.')
FLAGS = flags.FLAGS
```

(3)训练主函数:加载模型配置。

```python
@tf.contrib.framework.deprecated(None,'Use object_detection/model_main.py.')
def main(_):
  assert FLAGS.train_dir,''train_dir' is missing.'
  if FLAGS.task == 0:tf.gfile.MakeDirs(FLAGS.train_dir)
  if FLAGS.pipeline_config_path:
    configs = config_util.get_configs_from_pipeline_file(
        FLAGS.pipeline_config_path)
    if FLAGS.task == 0:
      tf.gfile.Copy(FLAGS.pipeline_config_path,
                    os.path.join(FLAGS.train_dir,'pipeline.config'),
                    overwrite = True)
  else:
    configs = config_util.get_configs_from_multiple_files(
        model_config_path = FLAGS.model_config_path,
        train_config_path = FLAGS.train_config_path,
```

```
      train_input_config_path = FLAGS.input_config_path)
    if FLAGS.task = = 0:
      for name,config in [('model.config',FLAGS.model_config_path),
                          ('train.config',FLAGS.train_config_path),
                          ('input.config',FLAGS.input_config_path)]:
        tf.gfile.Copy(config,os.path.join(FLAGS.train_dir,name),
                  overwrite = True)

  model_config = configs['model']
  train_config = configs['train_config']
  input_config = configs['train_input_config']
model_fn = functools.partial(
    model_builder.build,
    model_config = model_config,
    is_training = True)
```

(4)训练主函数:设计模型线程和迭代循环。

```
  def get_next(config):
    return dataset_builder.make_initializable_iterator(
        dataset_builder.build(config)).get_next()

  create_input_dict_fn = functools.partial(get_next,input_config)

  env = json.loads(os.environ.get('TF_CONFIG','{}'))
  cluster_data = env.get('cluster',None)
  cluster = tf.train.ClusterSpec(cluster_data) if cluster_data else None
  task_data = env.get('task',None) or {'type':'master','index':0}
  task_info = type('TaskSpec',(object,),task_data)

  ps_tasks = 0
  worker_replicas = 1
  worker_job_name = 'lonely_worker'
  task = 0
  is_chief = True
  master = ''

  if cluster_data and 'worker' in cluster_data:
    worker_replicas = len(cluster_data['worker']) + 1
  if cluster_data and 'ps' in cluster_data:
    ps_tasks = len(cluster_data['ps'])

  if worker_replicas >1 and ps_tasks < 1:
    raise ValueError('At least 1 ps task is needed for distributed training.')

  if worker_replicas > = 1 and ps_tasks > 0:
    server = tf.train.Server(tf.train.ClusterSpec(cluster),protocol = 'grpc',
                    job_name = task_info.type,
                    task_index = task_info.index)
    if task_info.type = = 'ps':
```

```
        server.join()
        return
    worker_job_name = '% s/task:% d' % (task_info.type,task_info.index)
    task = task_info.index
    is_chief = (task_info.type = = 'master')
    master = server.target
```

(5)训练主函数:记录训练日志,配置训练函数参数。

```
graph_rewriter_fn = None
    if 'graph_rewriter_config' in configs:
        graph_rewriter_fn = graph_rewriter_builder.build(
            configs['graph_rewriter_config'],is_training = True)

    trainer.train(
        create_input_dict_fn,
        model_fn,
        train_config,
        master,
        task,
        FLAGS.num_clones,
        worker_replicas,
        FLAGS.clone_on_cpu,
        ps_tasks,
        worker_job_name,
        is_chief,
        FLAGS.train_dir,
        graph_hook_fn = graph_rewriter_fn)
    print("模型训练完成!")

if _ _ name _ _ = = '_ _ main _ _':
    tf.app.run()
```

训练程序train.py的完整源代码文件参见U6-train.py.pdf。

完整源代码

train.py

步骤4 训练模型

运行训练程序train.py,如图6.14所示,程序读取配置文件Plate.config中定义的训练模型、训练参数、数据集,把训练日志和检查点保存到checkpoint目录中。

```
$ conda activate unit6
$ python train.py --logtostderr --train_dir checkpoint --pipeline_config_path data/plate.config
```

项目6 节能洗车房车牌识别

图6.14 训练模型

步骤5 可视化训练过程

在训练过程中打开tensorboard可以查看训练日志,如图6.15所示。训练日志中记录了模型分类损失、回归损失和总损失量的变化,通过Losses选项中的图表可以看到训练过程中的损失在不断变化,越到后面损失越小,说明模型对训练数据的拟合度越来越高。

*注意需要把地址修改成对应的数据处理服务器地址,然后在浏览器中输入对应地址和端口号查看。

```
$ tensorboard --host 172.16.33.11 --port 8889 --logdir checkpoint/
```

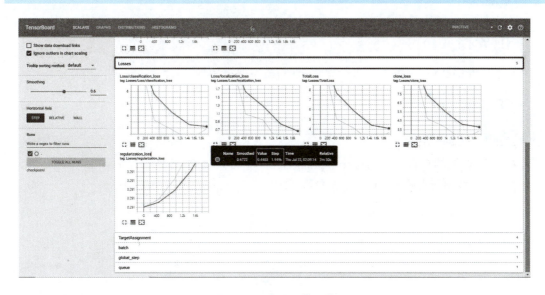

图6.15 查看训练日志

步骤6 查看训练结果

进入checkpoint子目录,可以看到生成了多组模型文件,如图6.16所示。其中:
- model.ckpt-xxxx.meta文件保存了计算图也就是神经网络的结构;
- model.ckpt-xxxx.data-xxxx文件保存了模型的变量;
- model.ckpt-xxxx.index文件保存了神经网络索引映射文件。

图 6.16 生成多组模型文件

任务 5　车牌识别模型评估

任务描述

在本任务中,将对训练模型进行评估,判断模型的可用性。

任务操作

步骤 1　创建评估程序

在开发环境中打开/home/student/projects/unit6/目录,创建评估程序 eval.py。

(1)导入模型所需模块,定义输入参数。

```
import functools
import os
import tensorflow as tf
from object_detection.builders import dataset_builder
from object_detection.builders import graph_rewriter_builder
from object_detection.builders import model_builder
from object_detection.legacy import evaluator
from object_detection.utils import config_util
from object_detection.utils import label_map_util

os.environ["TF_CPP_MIN_LOG_LEVEL"] = '3'
tf.compat.v1.logging.set_verbosity(tf.compat.v1.logging.ERROR)
flags = tf.app.flags
flags.DEFINE_boolean('eval_training_data',False,'')
flags.DEFINE_string('checkpoint_dir','','')
```

```
flags.DEFINE_string('eval_dir','','Directory to write eval summaries.')
flags.DEFINE_string('pipeline_config_path','','Path to a pipeline config.')
flags.DEFINE_string('eval_config_path','','')
flags.DEFINE_string('input_config_path','','')
flags.DEFINE_string('model_config_path','','')
flags.DEFINE_boolean('run_once',False,'')
FLAGS = flags.FLAGS
```

(2)评估主函数:加载模型配置。

```
@tf.contrib.framework.deprecated(None,'Use object_detection/model_main.py.')
def main(unused_argv):
  assert FLAGS.checkpoint_dir,''checkpoint_dir' is missing.'
  assert FLAGS.eval_dir,''eval_dir' is missing.'
  tf.gfile.MakeDirs(FLAGS.eval_dir)
  if FLAGS.pipeline_config_path:
    configs = config_util.get_configs_from_pipeline_file(
        FLAGS.pipeline_config_path)
    tf.gfile.Copy(
        FLAGS.pipeline_config_path,
        os.path.join(FLAGS.eval_dir,'pipeline.config'),
        overwrite = True)
  else:
    configs = config_util.get_configs_from_multiple_files(
        model_config_path = FLAGS.model_config_path,
        eval_config_path = FLAGS.eval_config_path,
        eval_input_config_path = FLAGS.input_config_path)
    for name,config in [('model.config',FLAGS.model_config_path),
                        ('eval.config',FLAGS.eval_config_path),
                        ('input.config',FLAGS.input_config_path)]:
      tf.gfile.Copy(config,os.path.join(FLAGS.eval_dir,name),overwrite = True)

  model_config = configs['model']
  eval_config = configs['eval_config']
  input_config = configs['eval_input_config']
  if FLAGS.eval_training_data:
    input_config = configs['train_input_config']

  model_fn = functools.partial(
      model_builder.build,model_config = model_config,is_training = False)
```

(3)评估主函数:定义评估循环,并记录评估日志。

```
  def get_next(config):
    return dataset_builder.make_initializable_iterator(
        dataset_builder.build(config)).get_next()

  create_input_dict_fn = functools.partial(get_next,input_config)

  categories = label_map_util.create_categories_from_labelmap(
```

```
        input_config.label_map_path)

    if FLAGS.run_once:
        eval_config.max_evals = 1

    graph_rewriter_fn = None
    if 'graph_rewriter_config' in configs:
        graph_rewriter_fn = graph_rewriter_builder.build(
            configs['graph_rewriter_config'], is_training = False)
```

(4)评估主函数:配置评估函数参数。

```
    evaluator.evaluate(
        create_input_dict_fn,
        model_fn,
        eval_config,
        categories,
        FLAGS.checkpoint_dir,
        FLAGS.eval_dir,
        graph_hook_fn = graph_rewriter_fn)
    print("模型评估完成!")

if __name__ == '__main__':
    tf.app.run()
```

评估程序 eval.py 的完整源代码文件参见 U6-eval.py.pdf。

完整源代码

eval.py

步骤2 评估模型

运行评估程序 eval.py,如图 6.17 所示,程序读取配置文件 Plate.config 中定义的训练模型、训练参数、数据集,读取 checkpoint 目录中的训练结果,把评估结果保存到 evaluation 目录中。

```
$ conda activate unit6
$ python eval.py --logtostderr --checkpoint_dir checkpoint --eval_dir evaluation
--pipeline_config_path data/plate.config
```

在评估过程中,可以看到对不同类别的评估结果。

图 6.17　评估模型

步骤 3　查看评估结果

利用 tensorboard 工具查看评估结果，如图 6.18 所示。

＊注意需要把地址修改成对应的数据处理服务器地址，然后在浏览器中输入对应地址和端口号查看。

```
$ tensorboard --host 172.16.33.11 --port 8889 --logdir evaluation/
```

步骤 4　分析模型可用性

在浏览器中查看各类别的平均精确度（AP）值，越接近 1 说明模型的可用性越高。此时图上显示 step 为 2k，说明该模型是训练到 2 000 步时保存下来的，对应 model.ckpt-2000 训练模型。

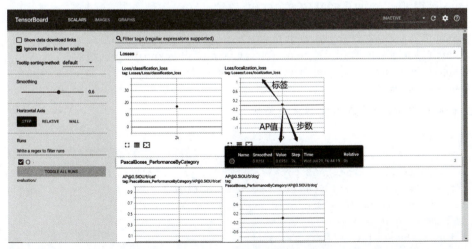

图 6.18　查看评估结果

任务6　车牌识别模型测试

任务描述

在本任务中,将把已经评估为可用性较强的模型,导出成为可测试的冻结图模型,用测试数据进行测试。

任务操作

步骤1　创建导出程序

在开发环境中打开/home/student/projects/unit6/目录,创建导出程序 export_fz.py。

(1)导入模型转换模块,定义输入参数。

```
import os
import tensorflow as tf
from google.protobuf import text_format
from object_detection import exporter
from object_detection.protos import pipeline_pb2

os.environ["TF_CPP_MIN_LOG_LEVEL"] = '3'
tf.compat.v1.logging.set_verbosity(tf.compat.v1.logging.ERROR)
slim = tf.contrib.slim
flags = tf.app.flags

flags.DEFINE_string('input_type','image_tensor','')
flags.DEFINE_string('input_shape',None,'[None,None,None,3]')
flags.DEFINE_string('pipeline_config_path',None,'Path to a pipeline config.')
flags.DEFINE_string('trained_checkpoint_prefix',None,'path/to/model.ckpt')
flags.DEFINE_string('output_directory',None,'Path to write outputs.')
flags.DEFINE_string('config_override','','')
flags.DEFINE_boolean('write_inference_graph',False,'')
tf.app.flags.mark_flag_as_required('pipeline_config_path')
tf.app.flags.mark_flag_as_required('trained_checkpoint_prefix')
tf.app.flags.mark_flag_as_required('output_directory')
FLAGS = flags.FLAGS
```

(2)转换模型主函数,调用模型转换函数。

```
def main(_):
  pipeline_config = pipeline_pb2.TrainEvalPipelineConfig()
  with tf.gfile.GFile(FLAGS.pipeline_config_path,'r') as f:
    text_format.Merge(f.read(),pipeline_config)
  text_format.Merge(FLAGS.config_override,pipeline_config)
  if FLAGS.input_shape:
    input_shape = [
        int(dim) if dim != '-1' else None
        for dim in FLAGS.input_shape.split(',')
```

```
    ]
else:
    input_shape = None
exporter.export_inference_graph(
    FLAGS.input_type,pipeline_config,FLAGS.trained_checkpoint_prefix,
    FLAGS.output_directory,input_shape = input_shape,
    write_inference_graph = FLAGS.write_inference_graph)
print("模型转换完成!")

if __name__ == '__main__':
    tf.app.run()
```

导出程序 export_fz.py 的完整源代码文件参见 U6-export_fz.py.pdf。

完整源代码

export_fz.py

步骤2 导出冻结图模型

运行导出程序 export_fz.py,如图 6.19 所示,程序将读取配置文件 Plate.config 中定义的配置,读取 checkpoint 目录中的 model.ckpt-2000 训练模型,导出为冻结图模型,并保存到 frozen_models 目录中。

```
$ conda activate unit6
$ python export_fz.py --input_type image_tensor --pipeline_config_path data/
plate.config --trained_checkpoint_prefix checkpoint/model.ckpt-2000 --output_directory
frozen_models
```

图 6.19 导出冻结图模型

步骤3　创建测试程序

在开发环境中打开/home/student/projects/unit6/目录,创建测试程序 detect.py。

(1)导入测试所需模块和可视化函数,定义输入参数。

```
import numpy as np
import os
import tensorflow as tf
import matplotlib.pyplot as plt
from PIL import Image
from object_detection.utils import label_map_util
from object_detection.utils import visualization_utils as vis_util
from object_detection.utils import ops as utils_ops

os.environ["TF_CPP_MIN_LOG_LEVEL"] = '3'
tf.compat.v1.logging.set_verbosity(tf.compat.v1.logging.ERROR)
detect_img = '/home/student/data/person/test/152.jpg'
result_img = '/home/student/projects/unit4/img/152_result.jpg'
MODEL_NAME = 'frozen_models'
PATH_TO_FROZEN_GRAPH = MODEL_NAME + '/frozen_inference_graph.pb'
PATH_TO_LABELS = 'data/label_map.pbtxt'
```

(2)加载模型计算图和数据标签。

```
detection_graph = tf.Graph()
with detection_graph.as_default():
    od_graph_def = tf.compat.v1.GraphDef()
    with tf.io.gfile.GFile(PATH_TO_FROZEN_GRAPH,'rb') as fid:
        serialized_graph = fid.read()
        od_graph_def.ParseFromString(serialized_graph)
        tf.import_graph_def(od_graph_def,name = '')
category_index = label_map_util.create_category_index_from_labelmap(PATH_TO_LABELS,
            use_display_name = True)
```

(3)图片数据转换函数。

```
def load_image_into_numpy_array(image):
    (im_width,im_height) = image.size
    return np.array(image.getdata()).reshape((im_height,im_width,3)).astype(np.uint8)
```

(4)单张图像检测函数。

```
def run_inference_for_single_image(image,graph):
    with graph.as_default():
        with tf.compat.v1.Session() as sess:
            ops = tf.compat.v1.get_default_graph().get_operations()
            all_tensor_names = {output.name for op in ops for output in op.outputs}
            tensor_dict = {}
            for key in ['num_detections','detection_boxes','detection_scores',
                'detection_classes','detection_masks']:
                tensor_name = key + ':0'
```

```
            if tensor_name in all_tensor_names:
                tensor_dict[key] = tf.compat.v1.get_default_graph().get_tensor_by_name
(tensor_name)
            if 'detection_masks' in tensor_dict:
                detection_boxes = tf.squeeze(tensor_dict['detection_boxes'],[0])
                detection_masks = tf.squeeze(tensor_dict['detection_masks'],[0])
                real_num_detection = tf.cast(tensor_dict['num_detections'][0],tf.int32)
                detection_boxes = tf.slice(detection_boxes,[0,0],[real_num_detection,-1])
                detection_masks = tf.slice(detection_masks,[0,0,0],[real_num_detection,-1,-1])
                detection_masks_reframed = utils_ops.reframe_box_masks_to_image_masks(
                    detection_masks,detection_boxes,image.shape[1],image.shape[2])
                detection_masks_reframed = tf.cast(tf.greater(detection_masks_reframed,0.5),
tf.uint8)
                tensor_dict['detection_masks'] = tf.expand_dims(detection_masks_reframed,0)
            image_tensor = tf.compat.v1.get_default_graph().get_tensor_by_name('image_
tensor:0')

            output_dict = sess.run(tensor_dict,feed_dict = {image_tensor:image})

            output_dict['num_detections'] = int(output_dict['num_detections'][0])
            output_dict['detection_classes'] = output_dict['detection_classes'][0]
.astype(np.int64)
            output_dict['detection_boxes'] = output_dict['detection_boxes'][0]
            output_dict['detection_scores'] = output_dict['detection_scores'][0]
            if 'detection_masks' in output_dict:
                output_dict['detection_masks'] = output_dict['detection_masks'][0]
    return output_dict
```

(5)输入图片数据,检测输入数据,保存检测结果图。

```
image = Image.open(detect_img)
image_np = load_image_into_numpy_array(image)
#[1,None,None,3]
image_np_expanded = np.expand_dims(image_np,axis = 0)
output_dict = run_inference_for_single_image(image_np_expanded,detection_graph)
vis_util.visualize_boxes_and_labels_on_image_array(
    image_np,
    output_dict['detection_boxes'],
    output_dict['detection_classes'],
    output_dict['detection_scores'],
    category_index,
    instance_masks = output_dict.get('detection_masks'),
    use_normalized_coordinates = True,
    line_thickness = 6)
plt.figure()
plt.axis('off')
plt.imshow(image_np)
plt.savefig(result_img,bbox_inches = 'tight',pad_inches = 0)
print("测试%s完成,结果保存在%s" %(detect_img,result_img))
```

测试程序 detect.py 的完整源代码文件参见 U6-detect.py.pdf。

detect.py

步骤 4　测试并查看结果

创建 img 目录存放测试结果,运行测试程序 detect.py,如图 6.20 所示,并查看输出的结果图片,如图 6.21 所示。

```
$ mkdir img
$ python detect.py
```

图 6.20　测试模型

图 6.21　查看测试结果

任务 7　车牌识别模型部署

任务描述

在本任务中,将把经过测试确认可用的模型,转换成为应用部署支持的格式,然后将模型文件部署到边缘计算设备上,并实现人工智能应用的集成。

任务操作

步骤1　创建导出程序

在开发环境中打开/home/student/projects/unit6/目录，创建程序 export_pb.py。

(1) 导入模型转换模块，定义输入参数。

```python
import os
import tensorflow as tf
from google.protobuf import text_format
from object_detection import export_tflite_ssd_graph_lib
from object_detection.protos import pipeline_pb2

os.environ["TF_CPP_MIN_LOG_LEVEL"] = '3'
tf.compat.v1.logging.set_verbosity(tf.compat.v1.logging.ERROR)
flags = tf.app.flags
flags.DEFINE_string('output_directory',None,'Path to write outputs. ')
flags.DEFINE_string('pipeline_config_path',None,'')
flags.DEFINE_string('trained_checkpoint_prefix',None,'Checkpoint prefix. ')
flags.DEFINE_integer('max_detections',10,'')
flags.DEFINE_integer('max_classes_per_detection',1,'')
flags.DEFINE_integer('detections_per_class',100,'')
flags.DEFINE_bool('add_postprocessing_op',True,'')
flags.DEFINE_bool('use_regular_nms',False,'')
flags.DEFINE_string('config_override','','')
FLAGS = flags.FLAGS
```

(2) 调用模型转换函数，完成模型转换。

```python
def main(argv):
  flags.mark_flag_as_required('output_directory')
  flags.mark_flag_as_required('pipeline_config_path')
  flags.mark_flag_as_required('trained_checkpoint_prefix')

  pipeline_config = pipeline_pb2.TrainEvalPipelineConfig()

  with tf.gfile.GFile(FLAGS.pipeline_config_path,'r') as f:
    text_format.Merge(f.read(),pipeline_config)
  text_format.Merge(FLAGS.config_override,pipeline_config)
  export_tflite_ssd_graph_lib.export_tflite_graph(
      pipeline_config,FLAGS.trained_checkpoint_prefix,FLAGS.output_directory,
      FLAGS.add_postprocessing_op,FLAGS.max_detections,
      FLAGS.max_classes_per_detection,FLAGS.use_regular_nms)
  print("模型转换完成!")

if __name__ == '__main__':
  tf.app.run(main)
```

导出程序 export_pb.py 的完整源代码文件参见 U6-export_pb.py.pdf。

export_pb.py

步骤2 导出pb文件

运行程序export_pb.py,读取配置文件plate.config中定义的参数,读取checkpoint目录中的训练结果,把tflite_pb模型图保存到tflite_models目录中。

```
$ conda activate unit6
$ python export_pb.py --pipeline_config_path data/plate.config --trained_checkpoint_prefix checkpoint/model.ckpt-2000 --output_directory tflite_models
```

步骤3 创建转换程序

在开发环境中打开/home/student/projects/unit6/目录,创建程序pb_to_tflite.py。

(1) 导入模块,定义输入参数。

```
import os
import tensorflow as tf
os.environ["TF_CPP_MIN_LOG_LEVEL"] = '3'
tf.compat.v1.logging.set_verbosity(tf.compat.v1.logging.ERROR)

flags = tf.app.flags
flags.DEFINE_string('pb_path','tflite_models/tflite_graph.pb','tflite pb file.')
flags.DEFINE_string('tflite_path','tflite_models/zy_ssd.tflite','output tflite.')
FLAGS = flags.FLAGS
```

(2) 转换为tflite模型。

```
def convert_pb_to_tflite(pb_path,tflite_path):
    # 模型输入节点
    input_tensor_name = ["normalized_input_image_tensor"]
    input_tensor_shape = {"normalized_input_image_tensor":[1,320,640,3]}
    # 模型输出节点
    classes_tensor_name = ['TFLite_Detection_PostProcess','TFLite_Detection_PostProcess:1','TFLite_Detection_PostProcess:2','TFLite_Detection_PostProcess:3']
    # 转换为tflite模型
    converter = tf.lite.TFLiteConverter.from_frozen_graph(pb_path,
                                                  input_tensor_name,
                                                  classes_tensor_name,
                                                  input_tensor_shape)

    converter.allow_custom_ops = True
    converter.optimizations = [tf.lite.Optimize.DEFAULT]
    tflite_model = converter.convert()
```

（3）tflite 模型写入。

```
converter.allow_custom_ops = True
converter.optimizations = [tf.lite.Optimize.DEFAULT]
tflite_model = converter.convert()
# 模型写入
if not tf.gfile.Exists(os.path.dirname(tflite_path)):
    tf.gfile.MakeDirs(os.path.dirname(tflite_path))
with open(tflite_path,"wb") as f:
    f.write(tflite_model)
print("Save tflite model at % s" % tflite_path)
    print("模型转换完成!")

if __name__ == '__main__':
    convert_pb_to_tflite(FLAGS.pb_path,FLAGS.tflite_path)
```

转换程序 pb_to_tflite.py 的完整源代码文件参见 U6-pb_to_tflite.py.pdf。

完整源代码

pb_to_tflite.py

步骤 4　转换 tflite 文件

运行程序 pb_to_tflite.py。

```
$ python pb_to_tflite.py
```

步骤 5　创建推理执行程序

在开发环境中打开/home/student/projects/unit6/tflite_models 目录,创建程序 func_detection_img.py。

（1）导入模块。

```
import os
import cv2
import numpy as np
import sys
import glob
import importlib.util
import base64
```

（2）定义模型和数据推理器。

```
def update_image(image_data,GRAPH_NAME = 'zy_ssd.tflite',min_conf_threshold = 0.5,
                use_TPU = False,model_dir = 'util'):
    from tflite_runtime.interpreter import Interpreter
    CWD_PATH = os.getcwd()
```

```
            PATH_TO_CKPT = os.path.join(CWD_PATH,model_dir,GRAPH_NAME)

            labels = ['plate','jing','A','B','C','D','E','F','G','H','J','K','L','M',
'N','P','Q','R','S','T','U','V','W','X','Y','Z','0','1','2','3','4','5','6','7',
'8','9']

            interpreter = Interpreter(model_path = PATH_TO_CKPT)

            interpreter.allocate_tensors()

            input_details = interpreter.get_input_details()
            output_details = interpreter.get_output_details()
            height = input_details[0]['shape'][1]
            width = input_details[0]['shape'][2]

            floating_model = (input_details[0]['dtype'] == np.float32)

            input_mean = 127.5
            input_std = 127.5
```

(3)输入图像并转换图像数据为张量。

```
# base64 解码
img_data = base64.b64decode(image_data)
# 转换为 np 数组
img_array = np.fromstring(img_data,np.uint8)
# 转换成 opencv 可用格式
image = cv2.imdecode(img_array,cv2.COLOR_RGB2BGR)

image_rgb = cv2.cvtColor(image,cv2.COLOR_BGR2RGB)
imH,imW,_ = image.shape
image_resized = cv2.resize(image_rgb,(width,height))
input_data = np.expand_dims(image_resized,axis = 0)

if floating_model:
    input_data = (np.float32(input_data)-input_mean)/input_std

interpreter.set_tensor(input_details[0]['index'],input_data)
interpreter.invoke()

boxes = interpreter.get_tensor(output_details[0]['index'])[0]
classes = interpreter.get_tensor(output_details[1]['index'])[0]
scores = interpreter.get_tensor(output_details[2]['index'])[0]
```

(4)检测图像,并可视化输出结果。

```
for i in range(len(scores)):
    if((scores[i] > min_conf_threshold) and (scores[i] < =1.0)):
        ymin = int(max(1,(boxes[i][0] * imH)))
```

```
            xmin = int(max(1,(boxes[i][1] * imW)))
            ymax = int(min(imH,(boxes[i][2] * imH)))
            xmax = int(min(imW,(boxes[i][3] * imW)))

            cv2.rectangle(image,(xmin,ymin),(xmax,ymax),(10,255,0),2)

            object_name = labels[int(classes[i])]
            label = '% s:% d% % ' % (object_name,int(scores[i] * 100))
            labelSize,baseLine = cv2.getTextSize(label,cv2.FONT_HERSHEY_SIMPLEX,0.7,2)
            label_ymin = max(ymin,labelSize[1] +10)
            cv2.rectangle(image,(xmin,label_ymin-labelSize[1]-10),(xmin + labelSize[0],
                label_ymin + baseLine-10),(255,255,255),cv2.FILLED)
            cv2.putText(image,label,(xmin,label_ymin-7),cv2.FONT_HERSHEY_SIMPLEX,0.7,
(0,0,0),2)

    image_bytes = cv2.imencode('.jpg',image)[1].tostring()
    image_base64 = base64.b64encode(image_bytes).decode()
    return image_base64
```

执行推理程序 func_detection_img.py 的完整源代码文件参见 U6-func_detection_img.py.pdf。

完整源代码

func_detection_img.py

步骤6　部署到边缘设备

把模型 zy_ssd.tflite 文件、推理执行程序 func_detection_img.py 文件复制到边缘计算设备中。

＊注意把 IP 地址修改成对应的推理机地址。

```
$ scp tflite_models/zy_ssd.tflite student@172.16.33.118:/home/student/zy-panel-check/util/
$ scp tflite_models/func_detection_img.py student@172.16.33.118:/home/student/zy-panel-check/util/
```

通过平台上的"应用部署"按钮,上传或输入图片 URL 检测,应用部署成功结果如图 6.22 所示。

人工智能图像识别项目实践

图6.22　应用部署成功

项目小结

在项目经理提供原始数据、资深算法工程师提供预训练模型的基础上，你顺利地完成了车牌识别的数据标注、模型训练，能够识别京牌车号，并部署到边缘计算设备上测试通过。如果没有团队的共同努力，很难在一周内完成这项工作。祝贺你和团队，为后续车牌识别功能的实际运用提供了可靠的基础。

多学一点：迁移学习是一种机器学习方法，就是把某一个任务中开发好的模型作为初始点，重新使用在另一个模型开发任务过程中。在计算机视觉任务中，利用迁移学习将预训练的模型作为新模型的起点是一种常用的方法。通常这些预训练的模型在开发神经网络的时候已经消耗了巨大的时间资源和计算资源，具备了一定的能力，迁移学习可以将这些技能迁移到相关领域的新问题上。按照目标领域有无标签分类，迁移学习可以分成监督迁移学习、半监督迁移学习、无监督迁移学习；按照学习方法分类，迁移学习可以分成基于样本的迁移、基于特征的迁移、基于模型的迁移、基于关系的迁移等。祝愿你在未来的学习中掌握更多的技能，在实际工作中灵活运用，成为一名优秀的工程师。